Practicals in POLYMER SCIENCE

Synthesis and Qualitative & Quantitative Analysis of Macromolecules

Practicals in Polymer Science

Synthesis and Qualitative & Quantitative Analysis of Polymers

provides a comprehensive discussion on synthesis, analysis and identification of polymers by simple physical and chemical methods. The book is designed to cater to the students of BE, BTech, MTech, BSc and MSc polymer courses as well as to the industrial needs. Graphic representation of the concepts are made wherever necessary and adequate theory is also dealt with. Many tables with properties of monomers, polymers, solvents and compounding ingredients are also included.

Dr. Siddaramaiah

is Assistant Professor in the Department of Polymer Science and Technology, SJ College of Engineering, Mysore. He obtained his MSc (chemistry) in 1986 and PhD in 1994 from University of Mysore. He has been actively involved in research on polymer materials for the past 15 years and has published more than 70 papers in international journals. He has worked in Central Leather Research Institute, Chennai, on SERC-DST Fellowship and has been a visiting scientist at IIT-Kharagpur in 2000.

Dr. Siddaramaiah is the recipient of *Young Scientist Award –1997* in chemical sciences from KAAS, Bangalore, and *Sir CV Raman Young Scientist Award –1999* in chemical technology from KSCST, Government of Karnataka. He worked at the Institute of Macromolecules, Federal University of Rio de Janeiro, Brazil, under TWAS-UNSCO Fellowship during 2005. He has successfully guided five scholars for PhD and at present a few more are pursuing PhD under him.

Practicals in POLYMER SCIENCE

Synthesis and Qualitative & Quantitative Analysis of Macromolecules

SIDDARAMAIAH
Department of Polymer Science and Technology
SJ College of Engineering
Mysore 570006
Karnataka
email : *siddaramaiah@yahoo.com*

CBS
CBS Publishers & Distributors Pvt. Ltd.
New Delhi • Bengaluru • Chennai • Kochi • Kolkata • Mumbai
Hyderabad • Uttarakhand • Nagpur • Patna • Pune • Jharkhand

ISBN: 81-239-1272-2

First Edition: 2005
Reprint: 2007, 2012, 2020

Published by **Satish Kumar Jain** and produced by **Varun Jain** for
CBS Publishers & Distributors Pvt. Ltd.,
4819/XI Prahlad Street, 24 Ansari Road, Daryaganj, New Delhi - 110002
delhi@cbspd.com, cbspubs@airtelmail.in • www.cbspd.com
Ph.: 23289259, 23266861, 23266867 • Fax: 011-23243014

Corporate Office: 204 FIE, Industrial Area, Patparganj, Delhi - 110 092
Ph: 49344934 • Fax: 011-49344935
E-mail: publishing@cbspd.com • publicity@cbspd.com

Branches:

- ***Bengaluru:*** 2975, 17th Cross, K.R. Road, Bansankari 2nd Stage, Bengaluru - 70 • Ph: +91-80-26771678/79 • Fax: +91-80-26771680
  E-mail: cbsbng@gmail.com, bangalore@cbspd.com
- ***Chennai:*** No. 7, Subbaraya Street, Shenoy Nagar, Chennai - 600030
  Ph: +91-44-26681266, 26680620 • Fax: +91-44-42032115
  E-mail: chennai@cbspd.com
- ***Kochi:*** Ashana House, 39/1904, A.M. Thomas Road, Valanjambalam, Ernakulum, Kochi • Ph: +91-484-4059061-65
  Fax: +91-484-4059065 • E-mail: cochin@cbspd.com
- ***Kolkata:*** 6-B, Ground Floor, Rameshwar Shaw Road, Kolkata - 700014
  Ph: +91-33-22891126/7/8 • E-mail: kolkata@cbspd.com
- ***Mumbai:*** 83-C, Dr. E. Moses Road, Worli, Mumbai - 400018
  Ph: +91-9833017933, 022-24902340/41 • E-mail: mumbai@cbspd.com

Representatives:

- Hyderabad: 0-9885175004
- Patna: 0-9334159340
- Jharkhand: 0-9811541605
- Nagpur: 0-9021734563
- Pune: 0-9623451994
- Uttarakhand: 0-9716462459

Printed at:
India Binding House, Noida, UP (India)

Preface

This is my privilege to design this book for the benefit of polymer science students, polymer engineering students (BE, MSc, MTech), polymer laboratory chemists and technologists, and R & D institution scientists. This book is broadly divided into three parts, namely: (i) fundamentals of polymer materials, (ii) synthesis of polymers, (iii) analysis and characterization of polymers by physical and chemical methods, and (iv) identification of polymers by qualitative method (simple physical and chemical methods). The polymer synthesis chapter also deals with the synthesis of both addition and condensation polymers. This book also gives to the students an insight into different polymerization methods such as bulk, solution, emulsion, and suspension polymerization.

This task was taken up with the sole intention of providing a comprehensive procedure for systematic qualitative and quantitative analysis of some polymers by physical and chemical methods before they approach the professionals or experts.

I am greatly indebted to His Holiness Jagadguru Sri Shivarathri Deshikendra Maha Swamigalavaru, President, JSS Mahavidyapeetha, Mysore, for the blessings. I express my indebtedness to Prof. K.S. Jagadeesh who was instrumental in nurturing professionalism in me. I sincerely thank Prof. M.H. Dhananjaya and Prof. B.S. Basavarajaiah for their moral support. My special thanks to Prof. B.G. Sangameshwara, Principal, and Dr. G.M. Shashidhara, Professor and Head, Department of Polymer Science and Technology, SJ College of Engineering, Mysore, for their encouragement throughout the preparation of this book.

I take this opportunity to thank Mr. K.R. Nan Kishore, Mr. K.H. Guruprasad, Mr. K. Shivamurthy and all those who helped me directly or indirectly to prepare this book. And, of course, I thank my parents Mr. Basavarajaiah and Smt. Rudramuniyamma, my wife Dr. Hemalatha S, and my daughter Kum.Chinmayee S, for the cooperation and support they extended at all times, most conspicuously during the writing of this book.

Any suggestions for improving this book will be highly appreciated.

Dr. Siddaramaiah

Contents

Preface v

Chapter 1
Fundamentals of Polymer Materials *1*

Introduction 1
Polymer 1
Classification of polymers 1
  Natural and synthetic polymers 1
  Organic and inorganic polymers 2
  Thermoplastics, thermosets and elastomers 2
  Homo-chain and hetero-chain polymers 3
  Homopolymers and copolymers 3
  Crystalline and amorphous polymers 3
  Classification of polymers depending upon the chain structure 4
Polymerization 4
Types of polymerization 4
Mechanism of polymerization 7
  Free radical polymerization 8
    Initiation step 8
    Propagation step 10
    Termination step 11
  Ionic polymerization 12
  Coordination polymerization 14
Methods of polymerization 16
  Bulk or mass polymerization 16
  Solution polymerization 17
  Emulsion polymerization 17
  Suspension polymerization 19
  Solid-state and gas phase polymerization 20
  Plasma polymerization 20
EXERCISE 23

Chapter 2
Synthesis of Polymers *24*

Introduction 24
EXPERIMENTS 25

01. Preparation of polystyrene by bulk polymerization method 25
02. Preparation of polystyrene by suspension (pearl or bead) polymerization method 27
03. Preparation of polystyrene by emulsion polymerization method 28
04. Preparation of polyacrylamide by free radical polymerization 30
05. Preparation of polyacrylamide by redox polymerization 32
06. Preparation of poly ethyl acrylate by emulsion polymerization 33
07. Precipitation polymerization of acrylonitrile 34
08. Suspension polymerization of methyl methacrylate 35
09. Radical co-polymerization of styrene and acrylonitrile 37
10. Preparation of polyaniline by free-radical polymerization 39
11. Preparation of phenol-formaldehyde resin in xylene medium 40
12. Preparation of urea-formaldehyde by polycondensation method 42
13. Preparation of aniline-formaldehyde resin by polycondensation method 43
14. Preparation of DGEBA epoxy resin using bisphenol-A and epichlorophydrine 44
15. Preparation of diglycidyl aniline epoxy resin 46
16. Preparation of polyester resin using ethylene glycol and maleic acid by polycondensation method 47
17. Preparation of nylon 6 from caprolactum 48
18. Preparation of nylon 6,6 from hexamethylene diamine and adipic acid 49
19. Preparation of cellulose secondary acetate from cellulose or bleached pulp 51
20. Preparation of cellulose triacetate from cellulose 52
21. Preparation of chlorosulphonated polyethylene from polyethylene. 53
22. Preparation of sulphonated polystyrene from polystyrene 54
23. Sheet casting using methyl methacrylate 55
24. Preparation of an emulsion paint using vinyl acetate 57
25. Preparation of alkyd resins 59
26. Preparation of polyvinyl alcohol from polyvinyl acetate (polymer modification) 60
27. Depolymerization of polymethyl methacrylate 61
EXERCISE 61

Chapter 3
Analysis and Characterization of Polymers 65

Introduction 65
EXPERIMENTS 66
01. Estimation of some physical properties of polymers 66
   (a) Determination of percentage of filler/fibre content in the polymer composite 66
   (b) Determination of ash content of the given polymer 67
   (c) Determination of moisture content of the polymer sample 67
   (d) Determination of water absorption of the polymer sample 68
   (e) Determination of melting point of the polymer sample 70
   (f) Determination of density of the polymer 70
   (g) Determination of bulk density for polymer powder or granules 72
02. Determination of rate of burning extent and time of burning of flexible plastic in a vertical position 75
03. Determination of viscosity average molecular weight and size of the molecule by viscometric method 78
04. Determination of acid value and iodine value for a given polymer sample 82
05. Determination of epoxy equivalent weight (EEW) by the hydrogen chloride-acetone method and hydrolyzable chlorine content of an epoxy resin 85
06. Determination of surface area of a carbon black sample by iodine absorption method 87
07. Determination of DBP value and Sieve analysis of carbon black 90
08. Determination of concentration of formaldehyde in formaline solution 92
09. Determination of aniline point and specific gravity of processing oil 93
10. Determination of concentration of hydrogen peroxide 96
11. Determination of the saponification value of a given oil 97
12. Determination of the flash, fire and pour point of a given oil 98
13. Determination of percentage of carbon black present in the polymer 103
14. Determination of viscosity of oil using Saybolt's and Redwood viscometers 104

15. Determination of percent extractable ingredients of a compounded rubber by solvent extraction method 107
16. Estimation of number average molecular weight by end group analysis 109
17. Estimation of the extent of swelling of polymer in different solvents 110
18. Estimation of free acid content, acetyl value and hydroxyl value of cellulose acetate 111
19. Determination of plasticizer in polyvinyl chloride (PVC) 116
20. Determination of network parameter of vulcanized rubber 118
21. Determination of nitrogen in a polymer using micro Kjeldahl apparatus (polymer sample - NBR or polyacrylamide) 120
22. Determination of chlorine content present in the given polymer by gravimetric method 122
23. Estimation of periodic acid induced degradation of polyvinyl alcohol 124
24. Determination of miscibility of polymer blends 125
    (a) Miscibility of polymer blend by viscosity method 126
    (b) Miscibility of polymer blend by density 127
    (c) Miscibility of polymer blend by refractive index method 128
    (d) Miscibility of polymer blend by ultrasonic velocity measurements 129
25. Estimation of resistance of plastic films to chemicals and to measure the weight change of films after immersion in chemicals as per ASTM D 1239-55 method 130
26. Determination of free sulphur and total sulphur from vulcanized rubber 131
27. Testing of natural rubber latex 134
    (a) Determination of total solid content (% TSC) 134
    (b) Determination of dry rubber content (% DRC) 135
    (c) Determination of mechanical stability 136
    (d) Determination of total alkalinity 136
    (e) Determination of KOH number 137
    (f) Determination of viscosity of latex by Brookfield viscometer 139
    (g) Determination of coagulum content 140
    (j) Determination of magnesium in latex 141

EXERCISE 145

Chapter 4
Identification of Polymers 149

Introduction 149
Rubbers 149
Flexible thermoplastics 153
Rigid thermoplastics 154
Thermosetting plastics 157

Part A : Identification of Plastics

I. Preliminary tests 160
II. Solubility test 179
III. Elemental analysis (tests for heteroatoms) 184
IV. Transition temperatures 189
V. Final identification by chemical tests 189
Specific identification tests 190
(a) Plastics containing nitrogen 190
(b) Plastics containing chlorine 197
(c) Plastics containing fluorine 199
(d) Plastics containing sulphur 202
(e) Plastics not containing nitrogen, sulphur and halogens 203
High temperature resistant thermoplastics 210
Miscellaneous properties of plastics 212

Part B : Identification of Rubbers

I. Preliminary tests 232
II. Solubility test 241
III. Elemental analysis 241
IV. Transition temperatures 242
V. Final identification by chemical tests 242
Group 1 - Rubbers containing nitrogen 242
Group 2 - Rubbers containing chlorine 243
Group 3 - Rubbers containing bromine 244
Group 4 - Rubbers containing fluorine 244
Group 5 - Rubbers containing Sulphur 244
Group 6 - Rubbers not containing nitrogen, halogens and sulphur 244
Miscellaneous properties of rubbers 249
EXERCISE 258

References 260

Appendices 263

Appendix I : Abbreviations for polymers for plastics and rubbers 265

Appendix II : Glossary of terms 271

Appendix III : Conversion tables 280

1

Fundamentals of Polymer Materials

INTRODUCTION

The science of macromolecules is divided into biological and non-biological materials. Biological polymers form the very foundation of life and intelligence and provide much of the food on which man exists. Non-biological polymers are the synthetic materials used for plastics, fibres, and elastomers with a few naturally occurring polymers like rubber, wool and cellulose. Today these substances are truly indispensable to mankind, being essential to clothing, shelter, transportation and communication as well as to the conveniences of modern living.

POLYMER

Polymer is a large molecule or macromolecule made up of many small molecules which combine to form a single long or large molecule. The individual small molecules from which the polymer is formed are known as *monomers*. Polymer can be of long-chain molecules or branched long-chain molecules or molecules of interconnected three-dimensional networks. The repeated unit of the polymer is equivalent or nearly equivalent to the monomer or starting material from which the polymer formed.

FICATION OF POLYMERS

Na ral and Synthetic Polymers

Based on their origin, polymers are classified into *natural* and *synthetic* polymers. Polymers obtained from natural sources are called *natural polymers*. Generally, natural polymers can degrade in nature and they are eco-friendly, for example, natural rubber, cotton, silk, wool, cellulose, etc. Cellophane, cellulose, rayon, leather, etc. are, infact, chemical modification of natural polymers. Polymers synthesized from low molecular weight materials (monomers) are called *synthetic polymers*. For example, polyolefins, nylon, polyester, synthetic rubber, epoxies, etc.

Organic and Inorganic Polymers

Polymers are also classified into organic and inorganic polymers. A polymer whose backbone (main chain) is essentially made-up of carbon atoms is termed an *organic polymer*. Organic polymers are formed by reacting organic monomers.

The molecules of inorganic polymer, on the other hand, generally contain no carbon atom in their backbone. Their main chain will contain non-carbon atoms like silicone rubbers, germanium, glass, borazine, polyphosphonitrile, etc. Inorganic polymers are thermally more stable compared to organic polymers.

Thermoplastics, Thermosets and Elastomers

Based on the effect of heat, polymers are classified into three groups, viz. thermoplastics, thermosets and elastomers.

Thermoplastics are linear or branched polymers which can be melted upon the application of heat. They are generally semi-crystalline in nature and soluble or swellable in solvents. They can be moulded and remoulded using conventional techniques. Examples are: PP, PMMA, PC, PVA, PAN, LDPE, PVC, PS, PPS, PEEK, PET, nylons, etc.

Thermosetting plastics or **thermosets** are heavily cross-linked polymers, which are normally rigid and intractable. They consist of a dense three-dimensional molecular network. They degrade rather than melt on the application of heat. Thermal and dimensional stability of thermosets are more than the thermoplastics. They do not melt and are insoluble in any solvents hence they are not recyclable. Examples are: Epoxies, PF, MF, unsaturated polyester resin, vulcanized rubber, etc.

Elastomers or rubbers display elastic properties, i.e. they can be stretched easily to high extensions and will spring back rapidly when the stress is released. Elastomers are soluble and recyclable in unvulcanized state and after vulcanization they become thermosets. Engineering properties of rubbers can be achieved after vulcanization. Examples are: NR, BR, EPDM, BIIR, NBR, etc.

Homo-chain and Hetero-chain Polymers

The polymers can be classified based on the nature of the element present in the polymer backbone. In homo-chain polymers, only one type of atoms (carbon–carbon) combine together to form the main chains. They are formed by the addition polymerization technique. Generally these are general purpose polymers, i.e. commodity plastics like HDPE, PP, PS, PMA, PVAc, PTFE, PVDC, PMMA, PVC, PVA.

In heteropolymers, more than one type of atoms combine together to form the main chain. They are formed by condensation polymerization technique. Most of the heteropolymers are engineering plastics, e.g. nylons, PET, PC, PEEK, PU, PPS, PPO, polyaramides, etc.

Homopolymers and Copolymers

Based on the type of monomer composition, polymers are classified into homopolymers and copolymers. Homopolymers are formed by polymerizing only one type of monomers whereas, polymerizing two or more types of monomers together forms copolymers. Examples for homopolymers are: PP, PS, PVC, PMMA, PC, PEEK, etc. Examples for copolymers are: SAN, SIS, ABS, NBR, SBR, EPDM, PS-g-MA, etc.

Copolymers are further classified into four different classes, depending upon the arrangements of the two monomers. They are: (i) *alternative copolymers*, (ii) *random copolymers*, (iii) *blocks copolymers* and (iv) *graft co-polymers*.

Crystalline and Amorphous Polymers

Thermoplastics are further divided into amorphous and crystalline polymers. Crystalline polymers are thermoplastic materials having long range well-ordered molecular structure. In these polymers, the chemical structures allow the polymer chains to fold on themselves and pack together in an organized manner. The resulting organized regions show the behaviour characteristics of crystals. Crystalline polymers possess higher melting point, density and mechanical performance compared to the corresponding amorphous polymers. Crystalline polymers are generally rigid, tough or brittle and dimensionally more stable. They are not soluble in solvents and are translucent or opaque in nature. These are generally characterized by crystalline melting point (T_m). Examples are: Nylon, PTFE, Isotactic PP, PET, HDPE, PAN, etc.

Amorphous polymers are thermoplastic materials having irregularly arranged molecular structure. When the molecules solidify they acquire random coil-like structure as in the molten state. Molecular arrangement will be very poor in these materials. Amorphous polymers are generally flexible, tough, and transparent in nature. These are soluble in solvents and are characterized by glass transition temperature (T_g). Examples are: PC, PS, PMMA, PVC, etc.

Classification Based on the Chain Structure

On the basis of chain structure the polymers are broadly classified as follows:

A. Flexible chains, cross-linked	• Vulcanized rubbers
B. Flexible chains, not cross-linked	• Flexible thermoplastics or unvulcanized rubbers
C. Rigid chains, not cross-linked	• Rigid thermoplastics or uncured thermosets
D. Rigid chains, cross-linked	• Thermosets

POLYMERIZATION

Polymerization is a process which allows simple low molecular weight compounds to combine and form a complex high molecular weight compound. For this each molecule of the compound should have the capability to react at least with two other molecules of the same or some other compound. In other words, they should have a functionality of at least two.

TYPES OF POLYMERIZATION

Carothers (1929) suggested classifying polymers and polymerization reactions as either condensation or addition. Correspondingly, two fundamental types of polymerization are termed as *addition polymerization* and *condensation polymerization.*

Addition polymers are usually formed because of a chain reaction through an active centre, whereas condensation polymers are formed by the process of step-wise intermolecular condensation of reactive functional groups. Therefore, Mark (1950) modified the nomenclature of the types of polymerization depending upon their reaction mechanism such as chain (addition or vinyl) polymerization and step (condensation) polymerization.

Chain (addition or vinyl) Polymerization

It is characterized by a self-addition of the monomer molecules, to each other, very rapidly through a chain reaction. No byproduct is formed; the product has same elemental composition as that of the monomer. Monomers or compounds containing reactive double (vinyl groups) or triple bonds can undergo chain polymerization reaction.

Step (condensation) Polymerization

Polycondensation is brought about by monomers containing two or more reactive functional groups (like hydroxyl, amino and carboxyl) condensing with each other. Step polymerization reactions are mostly accompanied by the elimination of small molecules. Compounds containing —OH, —COOH, —NH_2, etc. are the groups undergoing step polymerization reaction.

The differences between chain and step polymerization processesses are highlighted in Table 1.1.

Table 1.1: *Distinguishing mechanistic features of chain and step polymerization mechanisms*

Chain polymerization	*Step polymerization*
Only those molecular species containing active centres can add to the monomer units.	Any two potentially reactive molecular species present can react and combine.
Monomer concentration decreases throughout the reaction.	Rapid monomer depletion occurs very soon after the reaction begins.

(*contd.*)

Chain polymerization	*Step polymerization*
High molecular weight polymer is formed at once; polymer molecular weight changes little throughout the reaction.	Molecular weight increases steadily as the reaction progresses. A high degree of conversion is required for high molecular weight.
Long reaction times give high yields but affect molecular weight a little.	Long reaction times are essential to obtain high molecular weights.
Reaction mixture contains at anytime during the run only monomer, inactive high polymer and about ten parts active centres (growing chains).	At any stage of the reaction, all sizes of molecular species are present in a calculable distribution.
Examples: formation of polystyrene, PAN, PVA, PP, PMMA, PVC, polybutadiene, etc.	Examples: formation of polyesters, PC, UF, PF, PEEK, polyamides, polyurethanes (PUs), etc.

Table 1.2 indicates the functional groups of different monomers, and their reactions (condensation polymerization), polymer inter-unit linkage and the families of condensation polymers.

Table 1.2: *Types of condensation polymerization reaction polymers*

Reactants/ functional groups	*Polymer inter-unit linkage*	*Polymer type*	*Examples*
—OH + —COOH	$-C(=O)-O-$	Polyester	PET, PBT
$-NH_2$ + —COOH	$-N(H)-C(=O)-$	Polyamide (Nylons)	Nylon-6, Nylon-6, 6
—COOH + —COOH	$-C(=O)-O-C(=O)-$	Polyanhydride	PC

Reactants/ functional groups	*Polymer inter-unit linkage*	*Polymer type*	*Examples*
—NCO + —OH	—O—C(=O)—NH—	Polyurethane	PU, chain extended PU
—NCO + —COOH	—O—C(=O)—NH—	Polyurethane	- do -
—NCO + —NH_2	—NH—C(=O)—NH—	Polyurea	Polyurea
—OH + —OH	—O—	Polyether	PEG, Polyacetal
—CH—CH_2 (epoxide ring, —CH—CH_2— bridged by O)	—O—	Polyether	Epoxy
—OH + —COOR	—C(=O)—O—	Polyester	PET, PBT
—NH_2 + —C(=O)—NHR	—NH—C(=O)—NH—	Polyurea	Polyurea
—OH + —OR	—O—	Polyether or Polyacetal	Polyacetal, PEEK

MECHANISM OF POLYMERIZATION

There are three major steps in each polymerization process, which are: (i) initiation, (ii) propagation, and (iii) termination. The initiation of polymerization is made possible with the help of the initiator. Based on the nature of initiator, polymerization processes can be classified into three groups namely; free radical, ionic and co-ordination polymerization. Depending on the mechanism, there are three types of chain polymerization, as described in the forthcoming sections.

Free Radical Polymerization

Addition or chain reaction polymerization involves chain reaction in which the chain carrier may be an ion or a reactive substance with one unpaired electron is called a *free radical.* A free radical is usually formed by the decomposition of a relatively unstable material is called an *initiator.*

Initiation

In this group, the initiation is achieved by activating the monomer as a free radical. The use of high-energy sources will lead to the formation of monomer free radicals. However, this technique of using high energy sources will result in depolymerization of polymer chains. Therefore, the initiation step is generally carried out by adding low molecular weight compounds called *initiators.* They are thermally unstable compounds and are decomposed into free radicals which then activate the monomer molecules. Examples for these initiators are: azo-compounds, peroxides, hydroperoxide, peracids, and peresters. The decomposition of the initiator to form free-radicals (two fragments carrying one electron each) is called *homolytic decomposition.* The generalized reaction may be shown as:

$$R : R \rightarrow 2\ R^{\bullet}$$

Free radical can be induced by heat energy, light energy or catalysts. The illustration for the formation of free radical by each method is briefly highlighted below:

(a) *Thermal decomposition (decomposition by heat energy)*: The way thermal decomposition takes place can be illustrated by the following example of benzoyl peroxide;

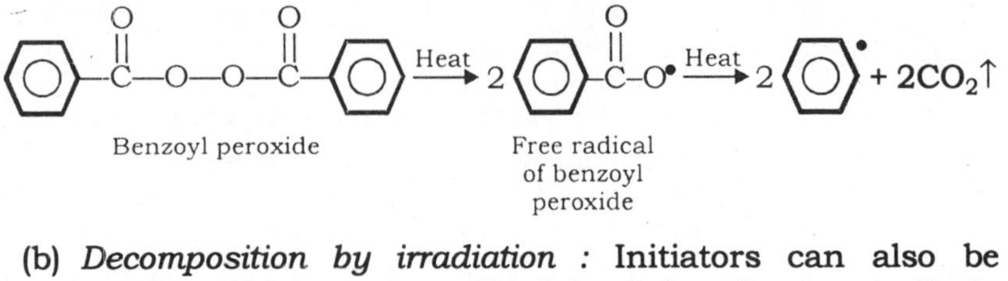

(b) *Decomposition by irradiation :* Initiators can also be decomposed by using ultraviolet light. The free radicals formed by the photodecomposition of an initiator are same as those formed by thermal decomposition.

$$(CH_3)_2\text{—}\underset{\displaystyle CN}{\underset{|}{C}}\text{—}N = N\text{—}\underset{\displaystyle CN}{\underset{|}{C}}\text{—}(CH_3)_2 \xrightarrow{h\nu} 2\,(CH_3)_2\text{—}\underset{\displaystyle CN}{\underset{|}{C^{\bullet}}} + N_2 \uparrow$$

2,2′-Azo bis isobutyro nitrile (AIBN) — Free radical of AIBN

where, hν represents light energy.

(c) *Decomposition using catalyst* : Initiators can also be induced to decompose into free radicals by using suitable catalysts. For example, benzoyl peroxide can be decomposed at room temperature with an aromatic tertiary amine.

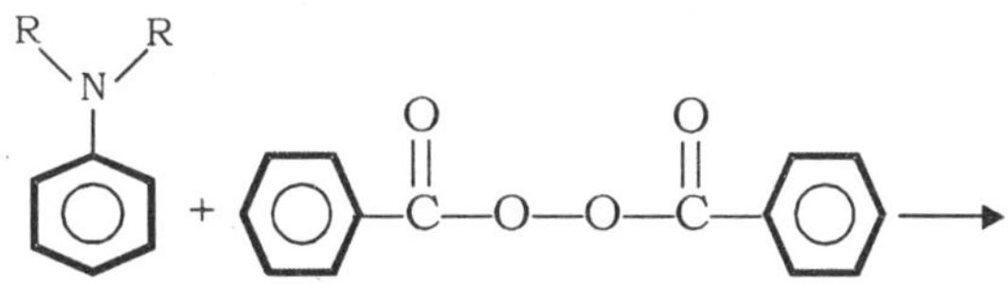

Aromatic tertiary amine — Benzoyl peroxide

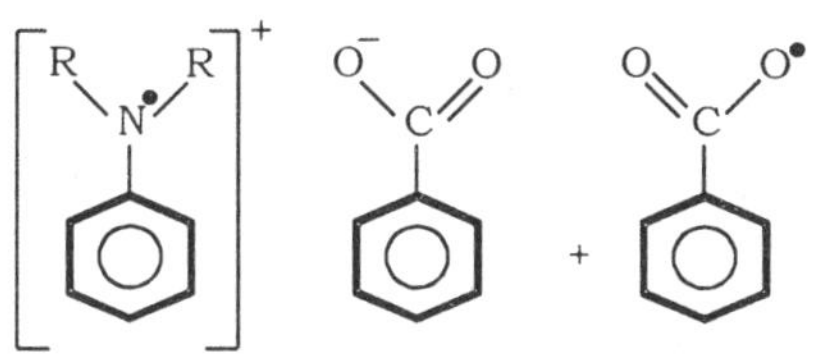

Free radical of initiator

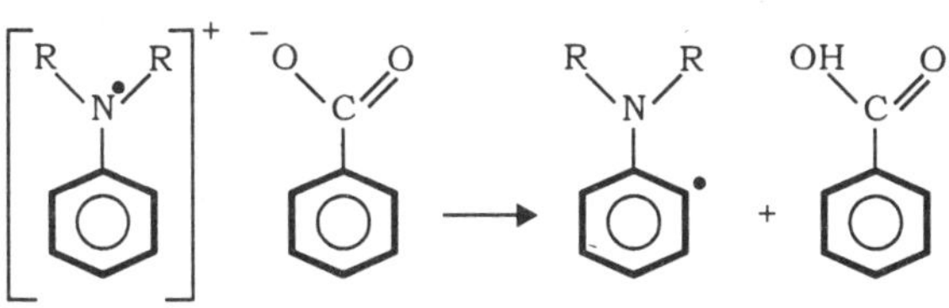

(d) *Redox initiators :* The decomposition of hydrogen peroxide by a ferrous ion and that of a hydroperoxide by a cobalt ion are the examples,

(i) H_2O_2 (Hydrogen peroxide) + Fe^{2+} (Ferrous ion) $\longrightarrow$ Fe^{3+} (Ferric ion) + OH^- + $HO^{\bullet}$ (Free radical)

(ii) $R\text{—}O\text{—}OH + Co^{2+} \longrightarrow RO^{\bullet} + Co^{3+} + OH^-$

(iii) $R\text{—}O\text{—}OH + Co^{3+} \longrightarrow ROO^{\bullet} + Co^{2+} + H^+$

Polymerization reactions utilizing such redox initiators are termed as *redox polymerization.* The rate of free-radical formation from initiators depend on the chemical nature, reaction temperature, solvent, intensity of heat and light radiation. Once the free radical is generated, the initiation takes place as follows:

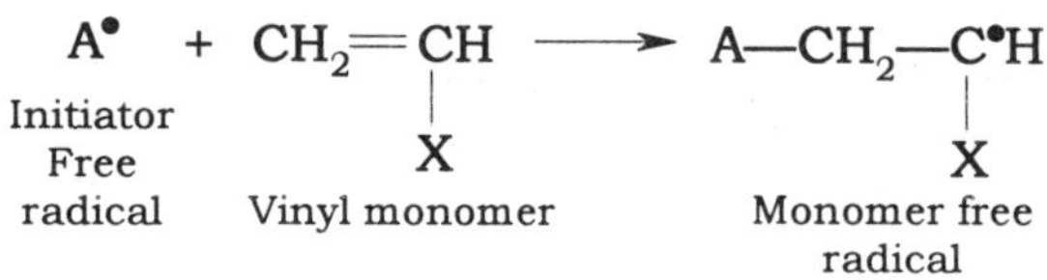

With this, the free radical site (active centre) is shifted from the initiator fragment to the monomer unit. During this shifting, the electrons of the double bond breaks-out forming a sigma (σ) bond with the initiator/monomer and retaining one electron on the active centre. This formation of a sigma-bond is generally accompanied by the liberation of energy. Hence, the free-radical attack on the monomer is an exothermic process, whereas, the formation of free-radical from decomposition of the initiator is an endothermic process.

Propagation step

In this step, the radical site (active centre) at the first monomer unit attacks the double bond of a fresh monomer molecule. This results in the linking-up of the second monomer unit to the first while shifting the radical site on to the second monomer.

This process involving a continuing attack on fresh monomer molecules each time. This process is called *propagation step.* A general representation of this process is as follows:

$$A_1^{\bullet} + M \longrightarrow A_2^{\bullet}$$
$$A_2^{\bullet} + M \longrightarrow A_3^{\bullet}$$
$$\bullet \quad \bullet \quad \bullet$$
$$\bullet \quad \bullet \quad \bullet$$
$$A^{\bullet}_{n-1} + M \longrightarrow A_n^{\bullet}$$

In the above vinyl addition, it may be seen that the head of the active chain adds to the tail of the monomer to produce a linear linkage. This is called *head-to-tail addition.* Tail-to-head, head-to-head, and tail-to-tail combinations are also possible.

Termination Step

The process of propagation continues as long as the monomer molecules are available for further addition. This is because of the presence of active species. If the kinetic chain-growth process is stopped by the annihilation arrest of the active species, then it is referred as *termination*. Termination can take place either:

(i) By collision between the growing chains at their active centres
(ii) By adding inhibitors or
(iii) By chain transfer.

By Collision

By collision, the probability of two growing chains coming close to each other predominates depending on the factors like temperature, time, monomer concentration, etc. This collision will then result in the arrest of chain-growth as follows:

(a) 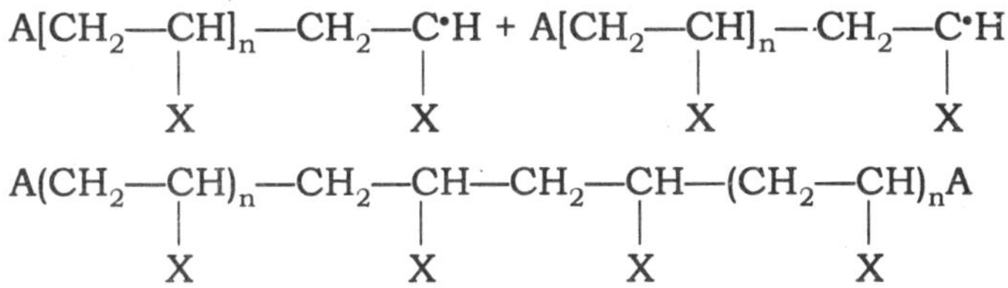

This type of coupling of lone pair of electrons is called *termination by coupling*. In this instance a long-chain (high molecular weight) polymer is obtained.

(b) 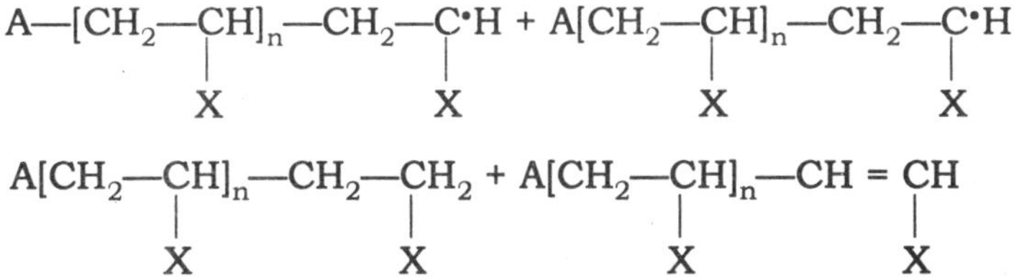

This type of termination is called *termination by disproportionation* resulting in the formation of two polymer molecules of shorter chain length (low molecular weight).

Inhibitors

In this case inhibitors are added to stop the growth of a chain. Inhibitors are substances that are capable of inhibiting or killing the chain growth by combining with active polymer free radicals and by forming either stable products or inactive free radicals. Therefore, a retarder or inhibitor is defined as a substance that can react with a free radical to form products incapable of adding monomer, e.g. hydroquinone, nitrobenzene, dinitrobenzene, benzothiazine, etc.

Atmospheric oxygen is a good inhibitor. The inhibiting action of oxygen is due to its biradical nature as depicted by:

$$\underset{\text{Polymer radical}}{\sim P^{\bullet}} + O^{\bullet}—O^{\bullet} \longrightarrow P—O—O^{\bullet}$$

$$P—O—O^{\bullet} + \sim P^{\bullet} \longrightarrow \underset{\text{Dead chain or polymer}}{P—O—O—P}$$

where, $\sim P^{\bullet}$ is the polymer radical. Because of the action of oxygen as inhibitor, generally the radical polymerization is carried out under inert atmosphere like nitrogen.

Chain Transfer

Termination takes place by the chain transfer. Here the activity is transferred from an initially active polymer chain to a previously inactive species. This means that while the growth of one polymer chain is stopped by forming a dead polymer, there is a simultaneous generation of a new free radical capable of initiating a fresh polymer chain growth. This reaction takes place by the abstraction of a hydrogen atom from the initiator, which may be depicted as:

$$A[CH_2—\underset{X}{\underset{|}{C}}H]_n—CH_2—\underset{X}{\underset{|}{C}}{}^{\bullet}H + TH \rightarrow A[CH_2—\underset{X}{\underset{|}{C}}H]_n—CH_2—\underset{X}{\underset{|}{C}}H_2 + T^{\bullet}$$

i.e., $AP_n^{\bullet} + TH \longrightarrow P_n + T^{\bullet}$

where, TH is the chain transfer species, P_n is the inactive polymer product and $T^{\bullet}$ is the newly created active species. The species TH, to which the reactivity is transferred may be monomer, solvent or a reagent deliberately added to promote chain transfer. Such species are referred to as a *chain transfer agent*. The major effect of chain transfer is to control the molecular weight and/or molecular weight distribution.

Ionic Polymerization

Chain polymerization can also proceed through the mechanism involving ions. Such reactions are called *ionic polymerization reactions*. In these reactions, the chain carriers are either *carbonium ions* or *carbanions*. Therefore, ionic polymerization reactions are classified into:

(a) Cationic polymerization (involving carbonium ions)
(b) Anionic polymerization (involving carbanions).

Cationic and anionic polymerization reactions are more complex and more versatile than the free radical polymerization reactions. These polar reactions are characterized by the rapid production of high molecular weight polymers at very low temperatures. Because of the ionic nature of reactions, the dielectric constant of the solvent is usually of overwhelming importance. Additional factors that are important are the solvating power of ions, temperature, monomer structure in terms of resonance, steric and polar effects.

Cationic polymerizations take place in vinyl monomers containing an electropositive substituent, e.g. styrene, vinyl-alkyl ether, alpha styrene, etc.

Ionic polymerization is a polymerization process in which the propagating species is a long chain cation or anion. Reactions are typically carried out in solvents of low or moderate polarity and generally proceed much faster than free radical polymerization. In contrast to radical polymerization, termination in ionic polymerization never involves the reaction between growing-chains, this makes the polymer molecular weight very high. The ionic polymerization involves three steps. They are as follows.

(i) *Formation of Initiator:* By ionization of the initiator, such as H^+ from BF_3 H_2O.

(ii) *Initiation:* Three cases of transfer of an electron from initiator to a monomer can be differentiated as following:

(a) A cation adds to a monomer to form a monomer cation. For example,

$$\underset{}{H^{\oplus}} + \underset{\text{Propylene}}{CH_2{=}CHCH_3} \longrightarrow \underset{\text{Propylene cation}}{CH_3C^{\oplus}HCH_3}$$

(b) An anion adds to a monomer to form a monomer anion. For example,

$$Bu^{\ominus}Li^{\oplus} + \underset{\text{Propylene}}{CH_2{=}CHCH_3} \longrightarrow \underset{\text{Propylene anion}}{Bu{-}CH_2{-}CH^{\ominus}CH_3Li^{\oplus}}$$

(c) A neutral molecule add, to a monomer to produce a zwitterion. For example,

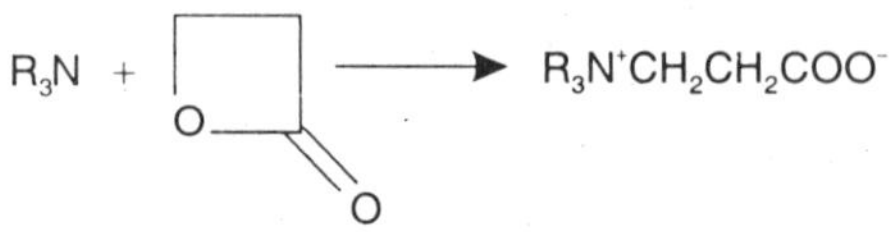

(iii) *Propagation and Termination:* The propagation reaction involves further adding up of the monomer units and simultaneous transfer of the charge to the newly added monomer unit. Termination occurs when a collision between the growing $C^{\oplus}$ ion and an anion (either from the counter-ion of another growing chain or from some deliberately added anion) takes place.

Unlike free radical polymerization, both cationic and anionic polymerization shows a marked dependence on the type of solvent used. High rate of polymerization at low temperature is a characteristic of ionic polymerization. The reaction normally occurs at low temperature. Ionic polymers are sometimes called as *Living polymers*, as all other species are eliminated, the growing chain remains active indefinitely. The polymer can have an extremely narrow distribution of molecular weight.

The polymer can be *killed* by addition of a terminating agent, such as water or other positive species or active hydrogen (hydrogen that can fall off the organic molecule) at the end of the reaction.

Catalysts for anionic polymerization processes include $NaNH_2$, BuLi, RMgX (Grignard reagent) and NaOH, all of which are very strong basic catalysts.

Cationic polymerization can be initiated by using many different chemicals, including Brownsted acids such as; BF_3 H_2O or $TiCl_4$ H_2O. Most cationic polymerizations appear to require the presence of a small molecule in trace amounts known as co-catalyst, like H_2O, HCl or RCl.

(Illustrations: $AlCl_3 + HCl \longrightarrow H^{\oplus} + AlCl_4$)

Coordination Polymerization

A unique polymerization technique using Ziegler or Ziegler-Natta catalyst. The product is normally a high linear chain and stereoregular polymer. Examples of Ziegler-Natta catalyst are: Triethyl aluminium (AlR_3) + Titanium tetra-chloride ($TiCl_4$), (diethyl aluminium chloride + vanadium trichloride) and (butyl

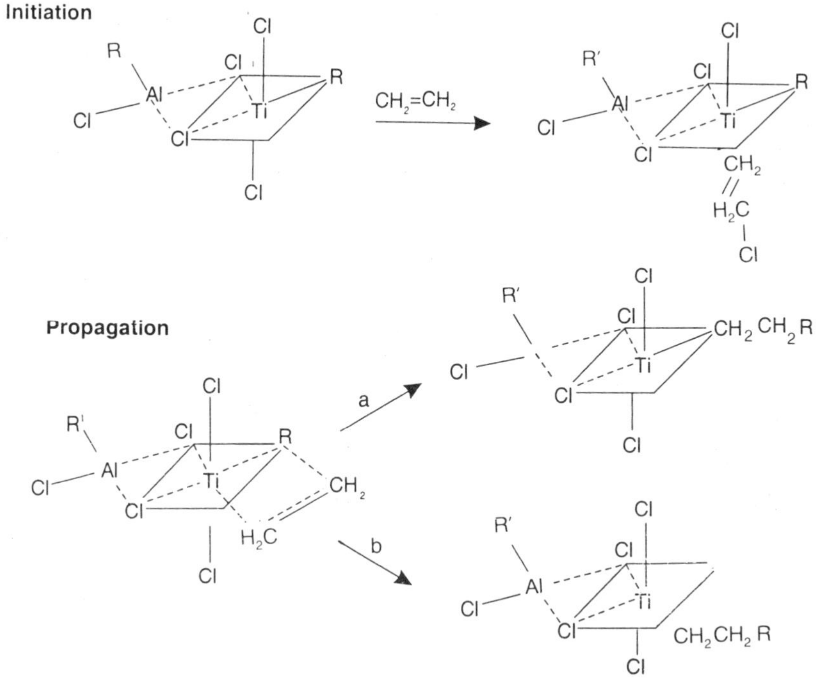

Fig. 1.1: *Initiation and propagation of coordination polymerization*

lithium + $TiCl_4$). In these catalysts, at least one empty coordination site accommodates the growing polymer chain. Under the Ziegler-Natta (ZN) process, ethylene is bubbled into the suspended catalyst at room temperature, and it rapidly polymerizes to a high molecular weight, linear polyethylene (HDPE). Linear polyethylene (as opposed to branched polyethylene) has a higher crystallinity (HDPE), so that the product differs significantly from polyethylene prepared from a high pressure (LDPE), a free radical process. The initiation, propagation, and termination steps for ZN polymerization are shown in the Fig. 1.1.

Diolefins (monomers with two double bonds) such as butadiene ($CH_2 = CH—CH = CH_2$) and isoprene yield both cis- and trans-1,4 polymers from Ziegler-Natta catalyzed processes. Variations in the metal halide and/or the ratio of alkyl to halide give different structures.

Termination step: The propagating end of the polymer chain is negatively charged and therefore, the reaction may be regarded as an anionic polymerization. Chain growth may be terminated by several different processes:

(i) Termination by spontaneous internal transfer (this destroys catalyst)
$Cat{-}CH_2{-}CHR{-}[{-}CH_2{-}CHR{-}]_n{-}CH_2CH_2R \longrightarrow Cat{-}H + CH_2{=}CR{-}[{-}CH_2{-}CHR{-}]_n{-}CH_2CH_2R$
In this reaction one fragment is picking up a hydrogen, and the other fragment is picking up a pi bond.

(ii) Termination by transfer with monomer (stops the reaction by adding a different olefin)
$Cat{-}CH_2{-}CHR{-}[{-}CH_2{-}CHR{-}]_n{-}CH_2CH_2R + CH_2{=}CHR \longrightarrow Cat{-}CH_2{-}CH_2R + CH_2{=}CR{-}[{-}CH_2{-}CHR{-}]_n{-}CH_2CH_2R$

(iii) A chain transfer type reaction may be brought about by the addition of an active hydrogen compound such as an alcohol:
$Cat{-}CH_2CHR{-}[{-}CH_2{-}CHR{-}]_n{-}CH_2CH_2R + R'{-}OH \longrightarrow Cat{-}OR' + CH_3{-}CHR{-}[CH_2{-}CHR{-}]_n{-}CH_2CH_2R$

Here, cat- denotes transition metals like Ti, Cr, V or Ni. The primed alkyl (R′) could be the same or different from the unprimed alkyl. The above chain transfer breaks a metal bond. In general, if an organometallic bond is broken, a proton goes to the carbon where the organometal was previously attached. A disadvantage of Ziegler-Natta catalysts is that the catalyst is dispersed in the final product.

METHODS OF POLYMERIZATION

Many polymerization processes are available in literature. Out of which, some commercially important methods are; bulk (mass), solution, emulsion, suspension (pearl or bead), solid-gas phase and plasma polymerization.

Bulk or Mass Polymerization

The monomer is taken in the liquid state and the initiator is dissolved in the monomer. The chain transfer agent, whenever used to control the molecular weight, is also dissolved in the monomer. Therefore, the synthesis is in a homogeneous phase. Bulk polymerization is a quite simple method and the product obtained has high purity since, except the initiator and the chain transfer agent, no other additive can contaminate the product. The polymer obtained can also be used as such since no isolation/purification from other components is involved. However, the disadvantages of this method are: (i) mixing becomes progressively difficult, because of drastic increase in viscosity, (ii) temperature control is not effective, (iii) termination becomes difficult and these polymers have a broad molecular weight distribution. The uncontrolled exothermic reaction can lead to an explosion.

Solution Polymerization

This process is also known as solvent polymerization. The monomer and initiator are dissolved in an inert solvent. Solution polymerization is carried out as a batch or continuous process. The advantage of this method is that the factors like viscosity, temperature and rate of polymerization can be controlled. The disadvantage of this method is that the removal or recovery of solvent is troublesome. Handling/disposing (toxicity) of solvent is very important. It can be used where the polymer is used in its solution form, as in the case of certain adhesives and coating compositions. The polymer formed is insoluble in its monomer or solvent and precipitate out and is amenable for easy separation.

Emulsion Polymerization

This process is extensively used by paint, adhesive and coating industries. The main recipe of this process is monomer, initiator (medium/solvent soluble), medium (generally aqueous), and emulsifying agent sodium laurel sulphate, examples— soap-like materials include common soaps, salts of carboxylic and sulphonic acids, etc. Besides these main process ingredients, other ingredients like buffer, co-monomer, catalyst, chain transferring agent, coagulant, etc., are required. In this process, high molecular weight product is obtained. The monomer is dispersed in the solvent (medium) in the form of droplets of about one micron (10^4 Å) size. Emulsifying agents stabilize these droplets, so that a thermodynamically stable emulsion (micelle) forms. Also, since the number of radicals in a micelle is very much limited and since there is a continous supply of monomer to the growing chain, a very high molecular weight product is achieved by this technique. The main disadvantages of this method are processing cost is very high and purification of polymer is difficult. The product formed in this process is in the form of emulsion. Formation of micelle is shown in the Fig. 1.2.

At the end of the polymerization, we have fine particles of the polymer, stabilized by the emulsifier layer and dispersed uniformly in an aqueous phase. This milky white dispersion is often called latex. The latex can either be used as such for making adhesives, water soluble emulsion paints, etc. The polymers can be isolated from the latex by destabilizing the emulsion (using some electrolyte) or by spray drying.

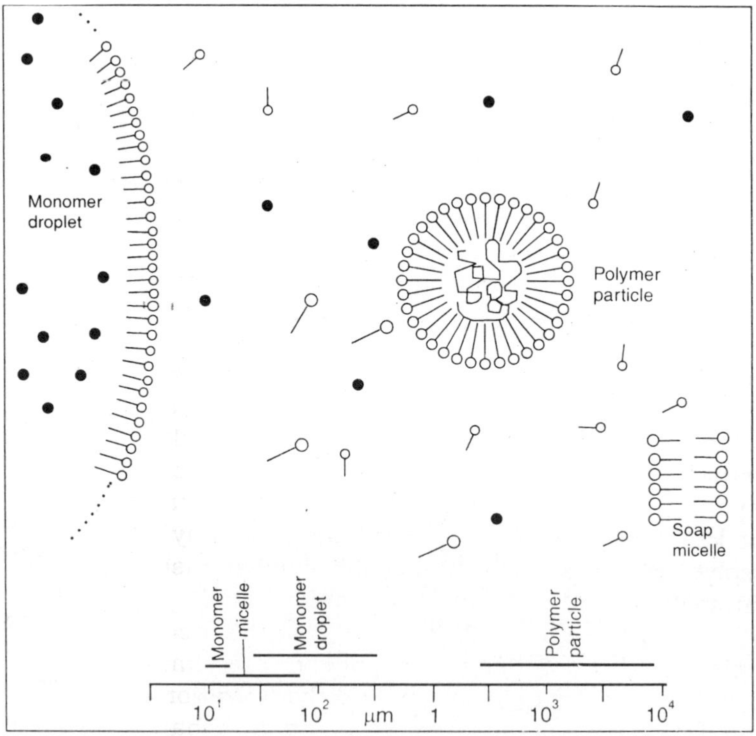

***Fig. 1.2:** Representation of the composition of an emulsion polymerization and relative particle size (Ref: P. Rempp and E. W. Merrill, Polymer Synthesis, Hüthig and Wepf, Basel, 1986).*

Suspension polymerization

This method was developed to overcome some of the disadvantages of the emulsion process, in particular, the tendency to absorb impurity due to large surface area of the small polymer particles. The recipe of this method is: monomer, initiator (monomer-soluble initiator is recommended, i.e., organic initiator), medium (solvent) and suspending agent. The usual dispersion medium is water. The product obtained by this method is in the form of bead or pearl. Hence, it is also referred as bead or pearl polymerization.

Suspension polymerization is used only in free radical type processes. The main requirement for suspension polymerization is monomer should be insoluble in solvent of media. The monomer is mechanically dispersed in a medium, usually water. There are cases where an organic medium is used in which neither the polymer nor the monomer is soluble. For suspension polymerization, there are two separate phases throughout the whole process, water and organic, and at the starting point may be 10 parts of the former, and 1 part of the latter. Usually monomer soluble initiaor (benzoyl peroxide or AIBN) have been, selected for this process. For suspension polymerization, initiation, propagation, and termination take place inside the droplet. The droplets must be kept far apart to keep the rate of suspension polymerization up with the rate of bulk polymerization, and to make the heat transfer much better. A protective colloid is added to increase the viscosity of the continuous phase so that it would hydrodynamically hinder the coalescence of monomer droplets (droplets coming together to form larger droplet is called coalescence). A suspending agent is a material that gives a surface activation that keeps droplets from coalescence. Polyvinyl alcohol (PVA) dissolved in the aqueous phase is a typical suspending agent.

Suspension polymerization is similar to bulk polymerization, and it could be considered "bulk polymerization within a droplet". The properties of the polymer are similar to those of the same polymer made by bulk polymerization. The polymer formed is also quiet good, resulting in a fairly narrow molecular weight distribution of the product. Isolation of the product becomes easy as this involves only filtration of the beads. The obtained product can be used as such for moulding purposes or can be dissolved in a suitable solvent for adhesives and coating applications. However, there are some limitations to this polymerization.

Limitations

Limitations of suspension polymerization are: it only applies to free radical process and agitation is critical because as the viscosity within the bead rises, the reaction rate increases suddenly (Tromsdoiff effect).

Solid-state and Gas Phase Polymerization

In addition to the usual methods of polymerization such as bulk and solution, polymers can be prepared in the solid state and gas or vapour phase. Solid phase polymerization is mostly restricted to chain polymerization. Generally, photo or radiation-activation technique is adopted in solid state polymerization. Since molecular mobility in the solid state is completely restricted, the polymer building process becomes quiet slow. Example: Solid state polymerization of trioxane into polyformaldehyde.

Gas phase polymerization is especially the case for the polymerization of olefins such as HDPE. In the Unipol process for PE, gaseous ethylene and solid catalyst (chromium or other complexes) are combined together in continuous fluidized-bed reactor. Since the polymerization is highly exothermic, proper management of heat transfer is critical to prevent agglomeration of the particles and to shutdown of the process. Variations of this process have been used for polymerization of propylene and co-polymerization of ethylene and propylene.

Plasma Polymerization

A wide variety of monomers can be polymerized in a plasma environment, which may consists of a low-pressure glow discharge of positively and negatively charged species, electrons, excited and neutral species, and electromagnetic radiation. Plasma polymerization can be used to prepare graft copolymers or to deposit a thin polymer film on metal or on other substrates (e.g. a corrosion-resistant coating on a metal). Plasma polymerization is a procedure, in which gaseous monomers, stimulated through plasma, condense on freely selectable substrates, as highly cross-linked layers. Condition for this process is the presence of chain-producing atoms, such as carbon, silicon or sulphur, in the working gas. As the monomer molecules in plasma, for the most part, shatter into reactive particles, there remain at most, only partially preserved chemical structures of the output gases in the product, which results in cross-linked and disordered structure as shown in Fig. 1.3. Structural preservation and cross-linking gradients can be controlled through process parameters, such as pressure, working gas-flow and applied electrical output; so that one can also construct so called gradient layers, i.e. with increasing degree of cross-linking over the thickness.

Out of this origination mechanism from plasma polymers, these result some special layer characteristics, which are qualified for a multitude of applications:

(i) Excellent coating adhesion on almost all substrates
(ii) Chemical, mechanical and thermal stability
(iii) High barrier effect.

Applications from plasma polymer coatings are: scratch-resistant coatings, corrosion protection, anti-bonding, anti-soiling coatings and barrier layers.

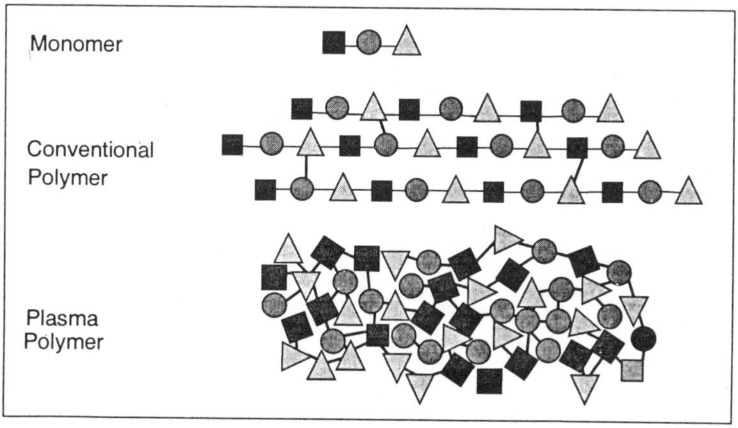

***Fig. 1.3:** Comparison of the structures of plasma polymers and conventional polymers*

Table 1.3: *Typical physical properties of some vinyl monomers*

Monomer	*Molecular Formula*	*M.W (g/mol)*	*Density (g/cc)*	*B. P. (°C)*	*R. I.*	*M. P. (°C)*	*Nature*
Styrene	$CH_2{=}CHC_6H_5$	104.15	0.909	145–146	1.547	–31	Cancer suspect and flammable liquid
Acrylonitrile	$CH_2{=}CHCN$	53.06	0.806	77	1.391	–83	Carcinogen
Acrylamide	$CH_2{=}CHCONH_2$	71.08	-		-	84–86	Cancer suspect agent
Acrylic acid	$CH_2{=}CHCO_2H$	72.06	1.051	139	1.420	13	Corrosive
Butyl methacrylate	$CH_2{=}C(CH_3){-}COOC_4H_9$	142	0.875	132	1.415	—	Flammable liquid
Vinylidene chloride	$CH_2{=}CCl_2$	97	1.213	30–32	1.425	–122	Cancer suspect and flammable liquid
4-Vinyl pyridene	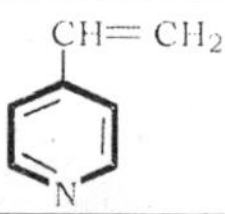	105.14	0.975	62–65	1.55		Corrosive, lachrymator
Methyl acrylate	$CH_2{=}CH(COOCH_3)$	86	0.956	79–81	1.402	–75	Flammable liquid
Methyl methacrylate	$CH_2{=}C(CH_3){-}COOCH_3$	100	0.936	100–101	1.414	86 (–48)	Flammable liquid and corrosive
Ehylene glycol	$HOCH_2{-}CH_2OH$	62.07	1.113	196–198	1.431	–13	Toxic and irritant

Ehylene glycol dimeth-acrylate	$[CH_2{=}C(CH_3)CO_2CH_2]_2$	188.22	1.051	98–100	1.454	–	Irritant
Aniline	$C_6H_5NH_2$	93.13	1.022	184	1.586	–6	Highly toxic and irritant
Formal-dehyde	HCHO	30	1.083	–	1.376	–	Cancer suspect agent

EXERCISE

1. What type of polymer structure will be formed by a monomer having more than two functional sites? What will be its characteristics?
2. How do linear and branched polymers differ from cross-linked polymers?
3. Outline the various techniques used for polymerization.
4. Define the following : (i) oligomers, (ii) high polymers, (iii) molecular weight distribution, (iv) crystallinity, (v) amorphous, (vi) polymerization, (vii) cross-linking reaction, (viii) tacticity, (ix) engineering plastics.

2 Synthesis of Polymers

INTRODUCTION

The preparation of polymers like polyacrylonitrile (PAN), polyacrylamide, polymethyl methacrylate (PMMA), polystyrene (PS), etc. through the process called polymerization is not a difficult task for the present-day polymer chemists. However, many of the polymers that we see today were totally unknown a few decades ago and indeed, synthetic polymer chemistry itself was a speculative, rather than a well-founded experimental science. To synthesize a few tonnes of polymers like polystyrene, all we need an appropriate quantity of the monomer and catalyst, and a suitable polymerization reactor. Moreover, many polymer-producing techniques have been so extensively studied and experimented upon, that we can almost always get a polymer of our choice in terms of the required molecular weight, structure, crystallinity, etc.

We have to repeat at this stage, for the sake of continuity, that polymerization is a process which allows simple low molecular weight compounds to combine and form a complex high molecular weight compounds. For this, each molecule of the compound should have the capability to react with atleast two other molecules of the same or some other compound/*monomer*. In other words, they should have a functionality of at least two. The functionality of a compound depends on the number of reactive sites it has. A compound assumes functionality because of the presence of reactive functional groups like —COOH, —OH, —NH_2, —COCl, etc. Some compounds, however, do not contain any reactive functional groups but the presence of double or triple bonds in the molecules bestows functionality to them. For example, ethylene, acetylene, butadiene, vinyl chloride, vinyl acetate, styrene, acrylonitrile, etc.

The presence of easily replaceable hydrogen atoms imparts functionality in some other compounds, e.g. phenol. Low molecular weight compounds having a functionality of two or more are called *monomers*. Depending on the functionality of the monomers used, we get linear, branched or three-dimensional crosslinked polymers.

EXPERIMENTS

Experiment 1

Preparation of Polystyrene by Bulk Polymerisation Method

Polystyrene (PS) can be prepared from styrene by using free radical, ionic or coordination polymerization methods. The functionality of styrene is two. It will undergo addition polymerization. Monomer soluble initiator should be used in bulk polymerization.

Bulk Polymerization

The monomer is taken in liquid state and the initiator is dissolved in the monomer. The reaction mass is heated or exposed to a radiation source for initiating the polymerization and is kept under agitation for proper mass and heat transfer. As the polymerization proceeds, the viscosity of the medium increases and mixing becomes progressively difficult, leading to polymers with very broad molecular weight distribution.

Reactions:

(*i*) $C_6H_5CO{-}O{-}O{-}OCC_6H_5 \longrightarrow 2C_6H_5COO^{\bullet}$

Benzoyl peroxide → Benzoyl peroxide free radical

(*ii*) $C_6H_5COO^{\bullet} + CH_2{=}CH(C_6H_5) \xrightarrow{\text{Initiation}} C_6H_5COO{-}CH_2{-}\dot{C}H(C_6H_5)$ (monomer-free radical)

(iii) $C_6H_5COO{-}CH_2{-}\dot{C}H(C_6H_5) + nCH_2{=}CH(C_6H_5) \xrightarrow{\text{Propagation}}$

$C_6H_5COO{-}[{-}CH_2{-}CH(C_6H_5){-}]_n{-}CH_2{-}\dot{C}H(C_6H_5){-}$

Termination

(iv) $C_6H_5COO{-}[-CH_2{-}CH(C_6H_5){-}]_nCH_2{-}\dot{C}H(C_6H_5) \rightarrow C_6H_5COO{-}(-CH_2{-}CH(C_6H_5))_n{-}CH{=}CH(C_6H_5)$

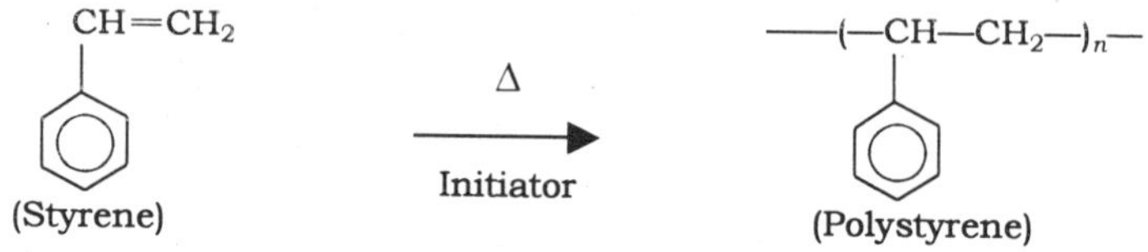

Chemicals: Styrene, azobis isobutyronitrile (AIBN) or benzoyl peroxide, dodecyl mercaptan (DDM).

Apparatus: 250 ml three-necked round-bottom flask, thermometer, hot plate cum magnetic stirrer (constant temperature bath), water-cooled condenser, vacuum desiccator, nitrogen gas cylinder.

Procedure: 50 mg of freshly recrystallised dry initiator (azobis iso butyro nitrile-AIBN or benzoyl peroxide) is taken in a clean and dry 250 ml round bottom flask. To this 10 ml of styrene is added. The flask is gently swirled until the initiator dissolves in the monomer. Another 10 ml of styrene and 500 mg of dodecyl mercaptan (DDM) is added to the flask and the contents are purged with oxygen free dry nitrogen gas for about 15 min. A water-cooled condenser is fitted to the flask and the flask is placed on a constant temperature bath maintained at 70 ± 1°C. The flask is swirled for 5 minutes. As the polymerization progress the viscosity of the substance increases. The increase in viscosity carefully observed and the flask is removed from the bath when the contents of the flask attain high viscosity. The contents are emptied into a 250ml beaker containing 25 ml of toluene. The flask is repeatedly washed with small quantity of toluene and washings are collected in the same beaker. The contents of the beaker are stirred with the magnetic stirrer until a homogeneous solution is formed. The polystyrene (PS) formed is now in the form of a solution in a mixture of toluene and the unpolymerized residual monomer. The polymer can be recovered by pouring the solution slowly into 500 ml of distilled methanol under constant agitation. The polymer will precipitate out as a white fluffy solid. The solid polymer is washed further with methanol and dried under vacuum at 60°C.

High purity product can be obtained by this method, because in this process we are using only monomer and less quantity of initiator. However, this process has many disadvantages such as controlling of temperature, viscosity, molecular weight distribution, etc.

Report: Report the yield and confirmation tests for polystyrene.

Experiment 2

Preparation of Polystyrene by Suspension (Pearl or bead) Polymerization Method

Principle: Only water-insoluble monomers can be polymerized by this technique. The monomer is suspended in water, in the form of fine droplets, which are stablised by using suitable water-soluble protective colloids, surface active agents and by stirring. The size of the monomer droplets formed depends on the monomer-to-water ratio, nature of stabilizing agents and also on the type and speed of agitation employed. The intiator should be monomer soluble. Polystyrene can also be prepared by suspension polymerization method, using monomer soluble initiator. Styrene monomer can be suspended by using water as media.

Reaction:

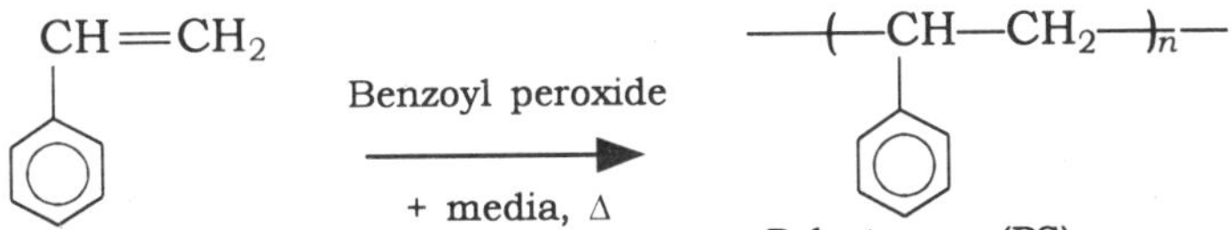

(Styrene) Polystyrene (PS)

Chemicals: Styrene, polyvinyl alcohol (PVA-suspending agent), benzoyl peroxide (initiator), medium (water), 1% NaOH solution.

Apparatus: 100 ml three necked round-bottom flask, thermometer, dropping funnel, nitrogen gas cylinder, hot plate-cum-magnetic stirrer and water-cooled condenser.

Procedure: About 30 ml of styrene monomer is freed from stabilizer by washing three times with 10 ml of 1% NaOH solution followed by three washings (20 ml each) with distilled water. The monomer is then dried over anhydrous $CaCl_2$.

A mixture of 50 ml of distilled water and 1g of PVA is taken in a 100 ml round-bottom flask and placed on a hot plate at 80 ± 2°C. To the washed 10 ml of styrene, 0.5 g of benzoyl peroxide (BPO) is added. The styrene containing benzoyl peroxide is added into the round-bottom flask by means of a dropping funnel over a period of 15 minutes with constant stirring. The reaction mixture in the water bath is stirred or shaken continuously for a period of 5 hour. The prepared polystyrene (PS) pearls or beads are filtered off, washed with 2% HCl and water, then dried. Note down the appearance and yield of PS.

Report: Report the yield, colour and confirmation tests for PS.

Experiment 3

Preparation of Polystyrene by Emulsion Polymerization Method

Principle: At the beginning of an emulsion polymerization, three phases can be differentiated. They are: a continuous aqueous phase containing the initiator; a phase of suspended droplets of monomer present in the aqueous phase not as discrete droplets, but as a uniform emulsion, and soap micelles containing small proportion of monomer. Initiator fragments enter some of the micelles, where polymerization takes place, supplied with monomer by diffusion from the monomer droplets through the aqueous phase. As these micelles (now properly called monomer-polymer particles) grow, they are stabilized by more soap at the expense of uninitiated micelles, which eventually disappear. The polymerization rate depends upon the number of micelles and this in turn on, the concentration of soap.

Reaction:

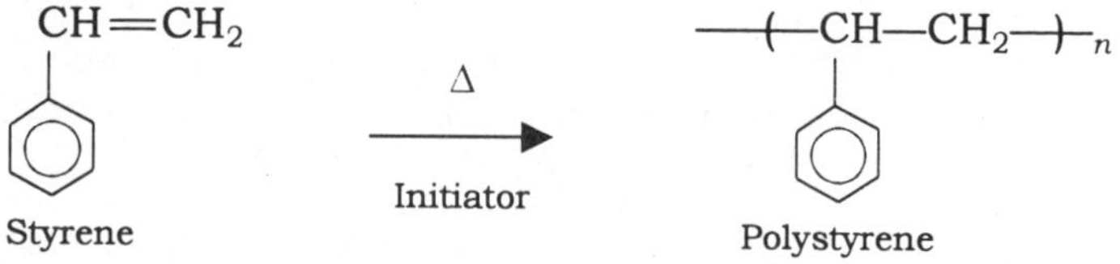

Chemicals: Styrene (monomer), sodium hydrogen phosphate (buffer solution), sodium lauryl sulphate (emulsifier), potassium persulphate (initiator), NaCl (coagulant), methanol and water (media/solvent).

Apparatus: 250 ml three-necked round-bottom flask (Quick fit flanged reaction flask), hot plate cum magnetic stirrer, nitrogen gas cylinder, water cooled condenser, vacuum desiccator, separating funnel and thermometer.

Special equipment: Quick fit flanged reaction flask equipped with stirrer, reflux condenser and separating funnel as shown in Fig. 2.1 and still-head unit is given in Fig. 2.2.

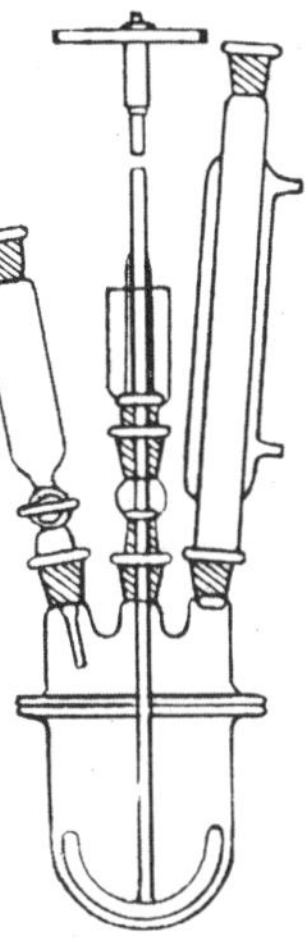

Fig. 2.1: *Flanged reaction flask with stirrer, reflux condenser and separating / tap funnel*

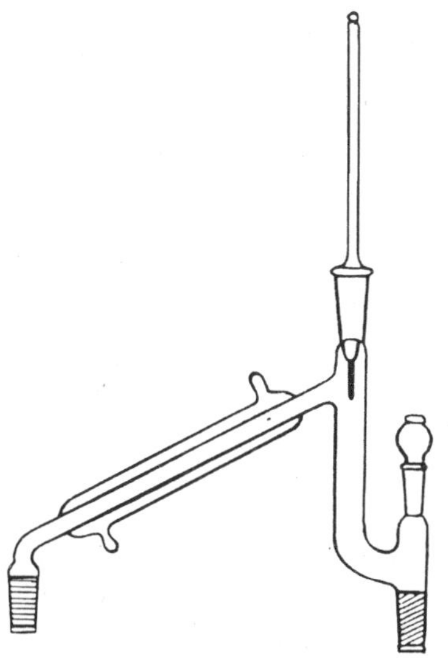

Fig. 2.2: *Still-head (Thermometer and reflux condenser)*

Procedure: 50 ml of distilled water is taken in a 250 ml three necked round bottom flask. To this, 10 mg of potassium persulphate (initiator), 10 mg of sodium hydrogen phosphate (buffer) and 0.5 g of sodium lauryl sulphate (emulsifier) are added. Further 10 ml of styrene monomer is added and the flask is fitted with a water-cooled condenser, a high-speed stirrer and an inlet tube dipping into the contents of the flask. Nitrogen gas is flushed slowly for a few minutes taking care that undue frothing do not take place. After flushing, the nitrogen inlet tube is removed and the mouth is closed with a stopper. The whole setup is placed in a constant temperature bath maintained at 70 $\pm$ 2°C and the mixture is vigorously stirred. If frothing makes the contents spill over the mouth of the flask, stirring should be adjusted. The experiment is continued for 4 to 6 hours, at the end of which the contents of the flask attain the form of a latex. The assembly is removed and the flask is taken out of the hot plate. To the latex in the flask, 50 ml of a 3% solution of aluminum sulphate or 3% NaCl in water is added slowly and stirred gently. When the emulsion is broken and the latex coagulates to give the polystyrene (PS) mass. The polymer is then washed with warm water and methanol and finally dried in vacuum at 60^0C.

Report: Report the yield and confirmation test for polystyrene (PS).

Experiment 4

Preparation of Polyacrylamide by Free Radical Polymerization

Principle: Acrylamide, is a crystalline solid melting at 84.5°C. The solid is quite stable, although ionizing radiation can cause polymerization. However, in bulk or in solution, the monomer is unstable at higher temperatures, especially if oxygen is completely absent. The polymerization is carried out conveniently in aqueous solution using a persulphate redox initiator at 50 – 70° C, or near room temperature when a reducing agent such as a bisulphate is added to accelerate the generation of free radicals. Inorganic persulphate like ammonium or potassium persulphate is used to initiate free radical polymerization of vinyl monomers in aqueous solution at 50–70° C or at room temperature.

The functionality of acrylamide is two. It also undergoes addition polymerization and the obtained polymer is in a gel state. It can be polymerized by using free radical initiator or redox initiator. It is a water-soluble polymer.

Reaction:

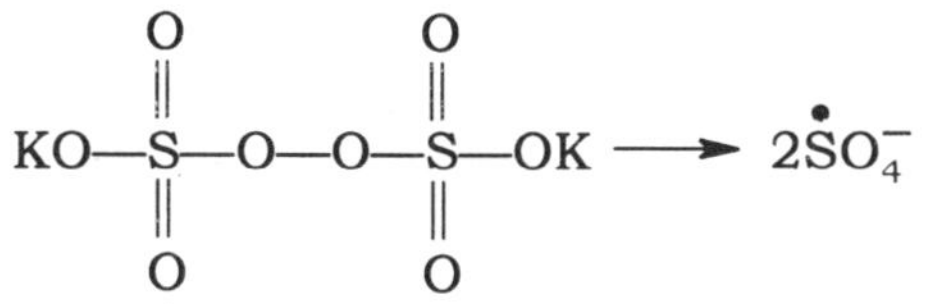

Potassium persulphate — Free radical of Initiator

$$\dot{S}O_4^- + H_2O \longrightarrow HSO_4^- + HO^{\bullet}$$

Sulphate free radical — Hydroxyl free radical

The hydroxyl free radical thus formed, adds to the vinyl monomer thereby initiating the polymerization.

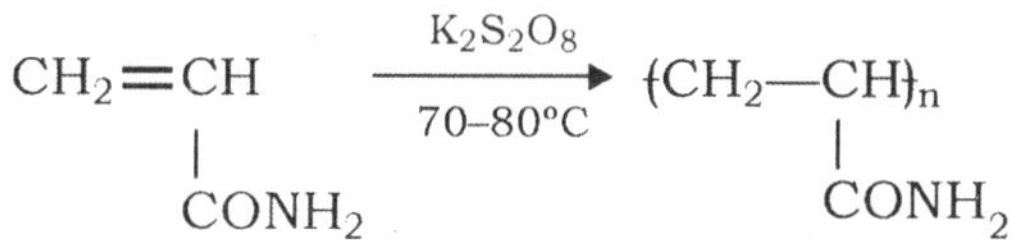

Acrylamide — Polyacrylamide

Chemicals: Acrylamide (monomer), potassium persulphate ($K_2S_2O_8$, initiator), methanol, medium (water).

Apparatus: 250 ml three-necked round bottom flask, hot plate cum magnetic stirrer, nitrogen gas cylinder, water-cooled condenser, vacuum desiccator, separating funnel and thermometer.

Procedure: Dissolve 5 g of acrylamide in 100 ml of water and place it in a 250 ml round bottom flask. Add 6 ml of 1 % solution of $K_2S_2O_8$ using separating funnel. The reaction mixture purged with nitrogen gas. Stirr well constantly using magnetic stirrer and gently heat for 30 minutes. One half of the reaction mixture is added drop wise to methanol and the precipitated polymer is dried in order to obtain the solid polymer. The obtained polyacrylamide product is white gel.

Report: Report the colour, yield, and confirmation tests for polyacrylamide.

Experiment 5

Preparation of Polyacrylamide by Redox Polymerization.

Principle: The acrylamide polymerization is carried out conveniently in aqueous solution using a redox initiator at room temperature. Ferrous ammonium sulphate in combination with hydrogen peroxide (Fenton's reagent) is used as the redox initiator, for the polymerization of acrylamide. The mechanism of initiation involves the formation of a hydroxyl ($HO^{\bullet}$) radical and ferric ion by the reaction of Fe^{2+} with H_2O_2 is given below. The radical so produced adds to the monomer and polymerization proceeds in the usual way.

Reaction:

Initiation step: The decomposition of hydrogen peroxide by a ferrous ion is as follows:

$$H_2O_2 + Fe^{2+} \longrightarrow Fe^{3+} + OH^- + HO^{\bullet}$$

Polymerization step:

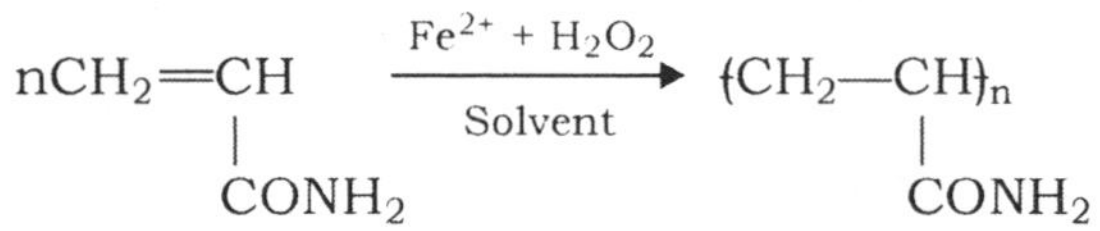

Acrylamide Polyacrylamide

In case of redox initiator system, the efficiency of the free radical production can be increased by the addition of a suitable reducing agent, which reduces the ferric ion back to ferrous ion.

Chemicals: Acrylamide (5 g), ferrous ammonium sulphate, hydrogen peroxide, methanol and water.

Apparatus: 250 ml three-necked round bottom flask, hot plate-cum-magnetic stirrer, nitrogen gas cylinder, water cooled condenser, vacuum desiccator, separating funnel and thermometer.

Procedure: Dissolve 10 g of pure acrylamide in 100 ml distilled water in a 250 ml flask. Purge the reaction mixture with oxygen free nitrogen gas. To this, add 25 ml of 0.5 M solution of ferrous ammonium sulphate in water and 25 ml of 0.1M solution of H_2O_2 in water with constant stirring for 30–40 minutes. The viscous solution is added with stirring to 250 ml of methanol containing 5 to 10 drops of HCl. The light brown coloured precipitate is filtered and dissolved in 50 ml of water. The solution is made ammonical for separation of iron from the polymer. The precipitated ferric hydroxide is filtered and the polymer solution is added drop wise from a dropping funnel with stirring to 100 ml of methanol. The precipitated polymer is allowed to settle for 10 minutes, filtered and dried in air. Note the appearance and yield. The polyacrylamide product obtained is white and hard.

Report: Report the colour, yield, and confirmation tests for polyacrylamide.

Experiment 6

Preparation of Poly Ethylacrylate by Emulsion Polymerisation

Principle: Ethylacrylate monomer is a liquid with boiling point of 99.5°C. It is commercially supplied with varying amounts of a polymerization inhibitor, 0.02 to 0.10 wt% hydroquinone being typical. The inhibitor level can be reduced to a negligible level by vacuum distillation or extraction by alkali. A redox couple is used to initiate the polymerization at room temperature.

Reaction:

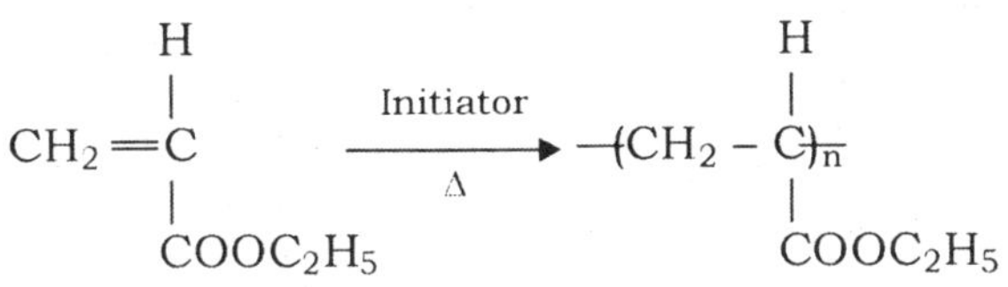

Ethyl acrylate Polyethyl acrylate

Chemicals: Ethyl acrylate, sodium lauryl sulphate, potassium persulphate (initiator), sodium meta bisulphate, ferrous sulphate, coagulate (NaCl, $Al_2(SO_4)_3$ or $CaCl_2$).

Apparatus: 250 ml three-necked round bottom flask, vacuum distillation equipment, separating funnel, thermometer, stirrer and Dewar flask.

Inhibitor removal:

1. Ethyl acrylate monomer is vacuum distilled to remove the inhibitor and stored in cooled condition, until furhter use. Do not pour ethyl acrylate down to the sink.
2. If vacuum distillation is not practical, wash 100 ml of monomer in a separating funnel with 50 ml of a 5% NaOH solution for 2 to 3 times. The monomer is then washed with water, successive 10 ml portion till this gives no alkaline reaction. The monomer is now allowed to stand over the granular calcium chloride till free of turbidity and moisture, and is then filtered into a clean conical flask and stored in cold.

Procedure:

1. Charge 100 ml of distilled water to the reactor and purge with nitrogen gas for 5 minutes with stirring.
2. Stop nitrogen flow and add 2 g solution of surfactant.
3. With continued stirring, add 10 ml of monomer.
4. Add 5 ml of 1 wt % of potassium persulphate. After few minutes, add 5 ml of 1 wt % solution of metabisulphate. After an additional 1 minute add 1ml of a 1 wt % solution of ferrous sulphate.
5. Record the temperature as a function of time at regular intervals of 10 minutes. After the temperature goes through a maximum arrest or terminate the polymerization reaction.
6. Coagulate 50 ml aliquots of the latex by the addition of 3% NaCl solution.
   (Examples for coagulants are; Methanol; 5% $CaCl_2 \cdot 2H_2O$; 5% NaCl or 5% $Al_2(SO_4)_3 18H_2O$)
7. After washing thoroughly with water, dry and calculate the total yield based on monomer charged.

Report: Report the colour, yield and confirmation tests of polyethyl acrylate (PEA).

Experiment 7

Precipitation Polymerization of Acrylonitrile

Principle: When the polymer is insoluble in its own monomer, a polymerization that begins with a homogeneous solution of monomer and initiator becomes heterogeneous as the polymer precipitates as a second phase. This is the case in the bulk or suspension polymerization of acrylonitrile, vinylchloride, etc. The kinetics of this polymerization is different from that of ordinary bulk polymerization. In the polymer phase, diffusion rates are so low, that termination is greatly slowed down. As the number of polymerization sites increases, both rate and molecular weight increase, until the concentration of monomer drops off at high conversion.

Reaction:

$$\underset{\text{Acrylonitrile}}{n\mathrm{CH_2{=}\underset{\displaystyle CN}{\underset{|}{C}H}}} \xrightarrow{\text{BPO}} \underset{\text{Polyacrylonitrile}}{\mathrm{-[-CH_2-\underset{\displaystyle CN}{\underset{|}{C}H}-]_{\mathit{n}}-}}$$

Chemicals: Acrylonitrile, benzoyl peroxide and toluene.

Apparatus: 250 ml three necked round bottom flask, constant temperature bath, stirrer, nitrogen gas cylinder, vacuum oven, separating funnel, thermometer, hot air oven.

Procedure: 50 mg of benzoyl peroxide is taken in 250 ml three necked round bottom flask and dissolved in 10 ml of acrylonitrile monomer and 100 ml of toluene is added to it. The contents in 250 ml round bottomed flask are flushed with nitrogen gas and the polymerization is started by keeping the flask in a constant temperature bath at 70 ± 2°C. After some time, turbidity develops and a white precipitate starts collecting at the bottom of the flask. In the case of acrylonitrile, the polymer formed is insoluble in its monomer as well as in toluene. Hence, the polyacrylonitrile (PAN) polymer precipitates as it is formed. The precipitated PAN is filtered from the liquid phase, washed several times with water and dried in vacuum at 60°C.

Report: Report the yield and confirmation tests of the PAN product.

Experiment 8

Suspension Polymerization of Methyl Methacrylate

Principle: Suspension polymerization, sometimes called *pearl* or *bead polymerization*, can be considered to be mini bulk polymerization carried out with the monomer dispersed in small droplets, typically 10 to 1000 μm in diameter. Droplets are suspended in a medium in which, the monomer is essentially insoluble (usually water), by means of careful, controlled agitation and with the aid of a dispersing agent or protective colloid. Dispersing agent is usually a water-soluble polymer such as polyvinyl alcohol (PVA) or methyl cellulose in the aqueous phase, but an oil soluble dispersing agent may also be added to the monomer phase. The dispersing agent prevents the aggregation of partially polymerized monomer particles at the stage when they are quite sticky. The size of the particles is controlled by the rate of agitation as well as amount and the nature of the dispersing agent used.

Reaction:

(*i*) $$\underset{\text{Benzoyl peroxide}}{C_6H_5CO{-}O{-}O{-}OCC_6H_5} \xrightarrow{\Delta} \underset{\text{Free Radical (R*)}}{2C_6H_5COO^{\bullet}}$$

(*ii*) $$C_6H_5COO^{\bullet} + CH_2{=}\overset{\overset{CH_3}{|}}{\underset{\underset{COOCH_3}{|}}{C}} \xrightarrow[\Delta,\ \text{Solvent}]{\text{Initiation}} \underset{\text{Monomer radical}}{C_6H_5COOCH_2{-}\overset{\overset{CH_3}{|}}{\underset{\underset{COOCH_3}{|}}{C^{\bullet}}}}$$

(*iii*) $$C_6H_5COOCH_2{-}\overset{\overset{CH_3}{|}}{\underset{\underset{COOCH_3}{|}}{C^{\bullet}}} + n\ CH_2{=}\overset{\overset{CH_3}{|}}{\underset{\underset{COOCH_3}{|}}{C}} \xrightarrow{\text{Propagation}}$$

$$C_6H_5COO{-}[{-}CH_2{-}\overset{\overset{CH_3}{|}}{\underset{\underset{COOCH_3}{|}}{C}}{-}]_n{-}CH_2{-}\overset{\overset{CH_3}{|}}{\underset{\underset{COOCH_3}{|}}{C^{\bullet}}} \quad (P^{\bullet})$$

(*iv*) $$\underset{\text{Polymer radical}}{P^{\bullet}} + \underset{\text{Initiator radical}}{R^{\bullet}} \xrightarrow{\text{Termination}} \underset{\text{Terminated polymer}}{P-R}$$

Polymer radical + Initiator radical ⟶ Terminated polymer

The overall reaction can be represented in simple form as follows.

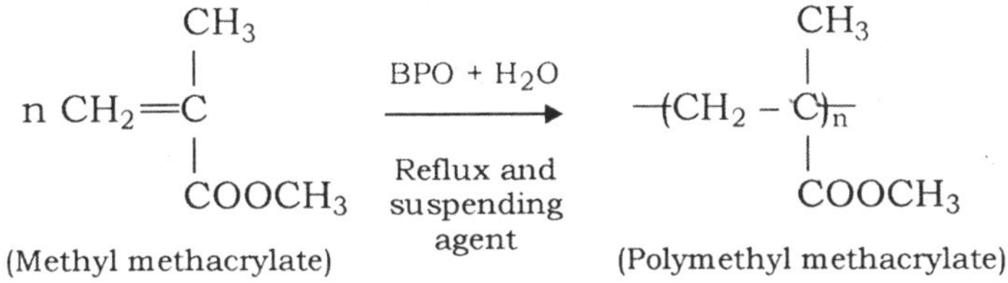

Chemicals: Methyl methacrylate, benzoyl peroxide, cetyl dimethyl benzyl ammonium chloride, polyvinyl alcohol (PVA), water (medium).

Apparatus: 250 ml three-necked round bottom flask, constant temperature bath, nitrogen gas cylinder, stirrer, water-cooled condenser and thermometer.

Procedure: 0.1g of benzoyl peroxide (initiator) is dissolved in 10 g of methyl methacrylate (monomer) contained in a 250 ml round bottom flask. To this, 40 ml of distilled water having 0.002 g of cetyl dimethyl benzyl ammonium chloride (or any other suitable quaternary ammonium salt type emulsifier) and 0.2 g of polyvinyl alcohol (protective colloid) are added. The contents of the flask are flushed with nitrogen gas and placed on a constant temperature bath maintained at 75 ± 1°C. A medium speed stirrer is fitted and the contents are stirred uniformly. It can be seen at this stage that the monomer is dispersed in water phase as tiny droplets. The size of the droplets can be adjusted by controlling the stirrer speed. The experiment is continued for 8 hrs. At the end of which, fine spherical beads of polymethyl methacrylate (PMMA) settles at the bottom of the flask. The beads are filtered and washed several times with warm water to free them from the emulsifier/ suspending agent.

Report: Report the nature, yield and confirmation tests for PMMA.

Experiment 9

Radical Co-Polymerization of Styrene and Acrylonitrile

Principle: Styrene and acrylonitrile form a copolymer not too far from random, with each type of growing chain-end preferring to add to the other monomer by about 2:1. The reactivity ratio expressed is determined from the composition of polymer formed at low conversion from known mixtures of the monomer.

Reaction:

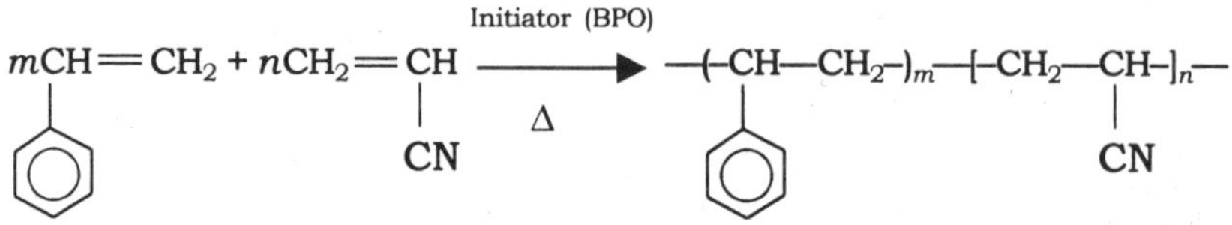

(Styrene) (Acrylonitrile) (Styrene-acrylonitrile copolymer)

Chemicals: Styrene (reagent grade, redistilled), acrylonitrile (reagent grade, redistilled), AIBN, benzoyl peroxide (BPO) (recrystalized), benzene, chloroform, hexane or methanol.

Apparatus: 5 large test tubes, constant temperature bath regulated at 70°C, 20 ml syringe, vacuum manifold and low pressure nitrogen supply, thermometer.

Procedure:

1. Weigh about 20 mg of benzoylperoxide each to four large test tubes and weigh 10 mg of AIBN in to a fifth tube. Cap the tubes with serum stoppers, evacuate them and flush with dry nitrogen gas.
2. Inject with the syringe, the following amounts of the monomers into the tubes; in the first four tubes, 4 ml acrylonitrile (AN) and 16 ml styrene (S); 8 ml AN + 12 ml S; 12 ml AN + 8 ml S; 14 ml AN + 6 ml S; and in the last tube, 10 ml AN.
3. Place the tubes protected with wire screens, on the constant temperature bath at 70°C and reflux them.
4. At the proper time, remove each tube from the bath and cool it quickly under the running water. Open the tube, and dilute the mixture with benzene or chloroform, if necessary to obtain fluidity.
5. Precipitate each styrene-acrylonitrile (SAN) copolymer quantitatively by pouring the polymerizate into about 300 ml petroleum ether, hexane or methanol. Collect the precipitate on a frilled-glass filter and dry to constant weight in a vacuum oven at 80°C. Determine the yield and see that it does not exceed 10% for the copolymers.

Report: Report the yield and confirmation tests for SAN.

Experiment 10

Preparation of Polyaniline by Free Radical Polymerization

Since the discovery of conducting polyacetylene in 1917, there has been continuing interest in scientific and engineering communities in understanding Intrinsically Conducting Polymers (ICPs) and finding applications for their unique properties. The most common ICPs of current interest are polyaniline (PAni), polyacetylene (PA), polypyrrole (PPy), polythiophene (PTh), poly *p*-phenylene (PPP) and poly *p*-phenylene sulphide (PPS).

Polyaniline has been the subject of intensive investigations during the last two decades because of its straightforward polymerization and desirable electrical, electrochemical and optical properties coupled with excellent environmental stability. Being inexpensive and relatively simple to synthesize, polyaniline has immense industrial potential for a number of applications such as in static films for transparent packaging of electronic components, anti-corrosion coatings for steels, electromagnetic shielding, rechargeable batteries, light emitting diodes (LEDs), nonlinear optical devices, sensors for medical and pharmaceutical purposes and membranes for separation of gas mixtures.

Principle: Aniline will undergo polymerization in the presence of ammonium persulphate in aqueous acidic medium. The colour of the polyaniline product is blue-black. Polyaniline can be easily prepared by using any free radical initiators. The yield and conductivity of the product depends upon the nature and amount of initiator and dopants used.

Reaction:

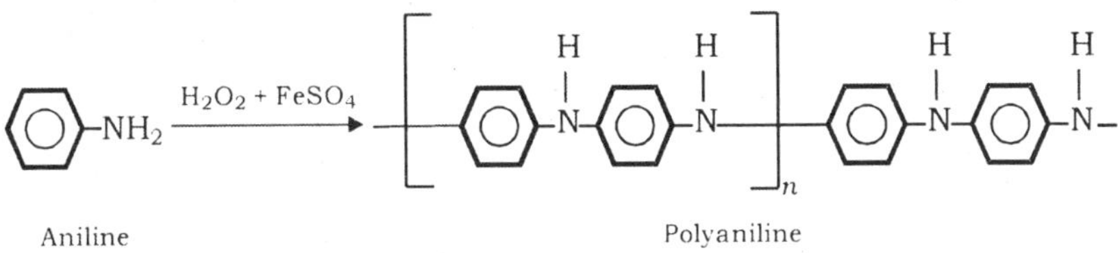

Aniline Polyaniline

Chemicals: Aniline, ammonium persulphate or benzoyl peroxide, 1M HCl or any other acid.

Apparatus: 250 ml beaker, hot plate-cum-magnetic stirrer and thermometer.

Procedure: 4.6 ml of aniline and 100 ml of aqueous hydrochloric acid (1M) is taken in 250 ml beaker. 100 ml of 0.1 M ammonium persulphate solution is added to the above solution through separating funnel with vigorous stirring over a period of 15 minutes. After the addition, the reaction mixture is continuously stirred for 10–12 hours. The reaction mixture is slowly poured into a beaker containing 50-100ml of acetone (non-solvent) with constant stirring. A blue-black powder of polyaniline is precipitated out. The powder is filtered and washed with water and with ethanol.

Report: Report the yield, colour and confirmation tests for polyaniline.

Experiment 11

Preparation of Phenol - Formaldehyde Resin in Xylene Medium

Principle: The polymeric products obtained by condensation polymerization of phenol with formaldehyde are called *phenol-formaldehyde* (PF) resins. These are mixtures of structurally polydisperse compounds which are initially soluble and meltable and which can be cross-linked (hardened) by subsequent reactions. Different end-products are obtained depending on whether the polycondensation is carried out with acidic or basic catalysts and also whether the phenol or formaldehyde is in excess. Condensation in acidic medium results in soluble, meltable phenol resin essentially linked via methylene groups in the *ortho-* and *para-* positions of the phenol and have average molecular weight of 600–1500; they are called *novolacs.* Prolonged heating of these products does not affect further condensation; they may however, be cross-linked with suitable reagents. The soluble novolac resins have melting points between 100 and 140°C.

The reactions are generally carried out in aqueous medium by slowly adding formaldehyde solution to an acidic phenol or by mixing the components at room temperature and then heating. The product is washed with hot water to remove the acid and then distilled in vacuum to remove all water and unconverted phenol. The resin obtained is generally soluble in alcohol, lower esters, ketones and dilute alkalis.

The functionality of phenol is three and it is variable depending upon the concentration of formaldehyde. Suppose the molar ratio of phenol:formaldehyde is 1:0.8, then functionality of phenol is 2. If the molar ratios between phenol:formaldehyde is 1:1.2, then functionality of phenol is 3. That means two grades of PF resin are available commercially. They are novolac and resol. *Novolac* is a thermoplastic PF resin and *resol* is a thermoset PF. Novolac can be obtained when P:F molar ratio is 1:0.8. It is linear and soluble polymer. When the P:F molar ratio is 1:1.2, the yield is resol. It is a cross-linked polymer and not remouldable.

Reaction:

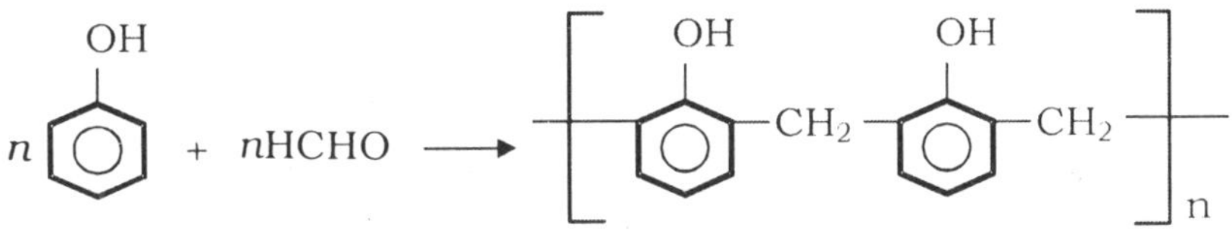

Phenol Formaldehyde Phenol-formaldehyde (PF)

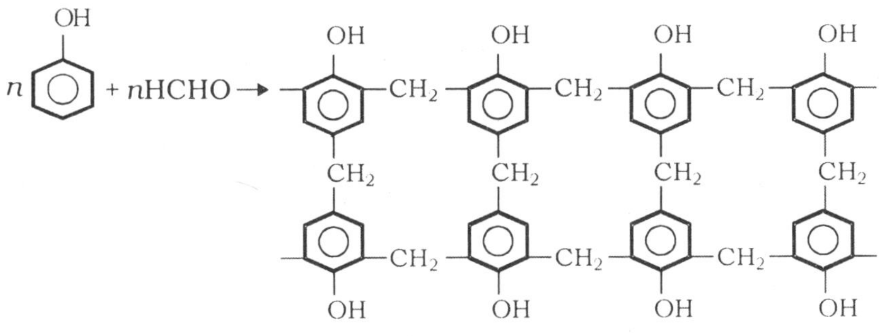

Chemicals: Phenol, formaldehyde (37% aqueous solution), oxalic acid and medium (water).

Apparatus: 500 ml three-necked round bottom flask, hot plate-cum-magnetic stirrer, thermometer and water-cooled condenser.

Procedure: 500 ml of three-necked round bottom flask equipped with stirrer and reflux condenser is charged with 65 g (1.38 moles) of phenol, 46 g of 37% aqueous formaldehyde solution (1.13 moles) and 2 g of oxalic acid dihydrate in 15 ml of water. The mixture is refluxed for one and half hours. After addition of 300 ml of water, the mixture is agitated thoroughly and is allowed to cool off. The aqueous layer is drawn off and the remaining water is removed by slowly raising the temperature to 50°C, and applying a vacuum by means of a water pump. This temperature is maintained until samples taken from the melt, which solidifies on cooling to form a rigid, brittle solid.

Report: Report the yield, colour and confirmation tests for Phenol-Formaldehyde resin.

Experiment 12

Preparation of Urea-Formaldehyde by Polycondensation Method

Principle: Formaldehyde reacts with urea in alkaline or acidic medium yielding methylol urea.

Reaction:

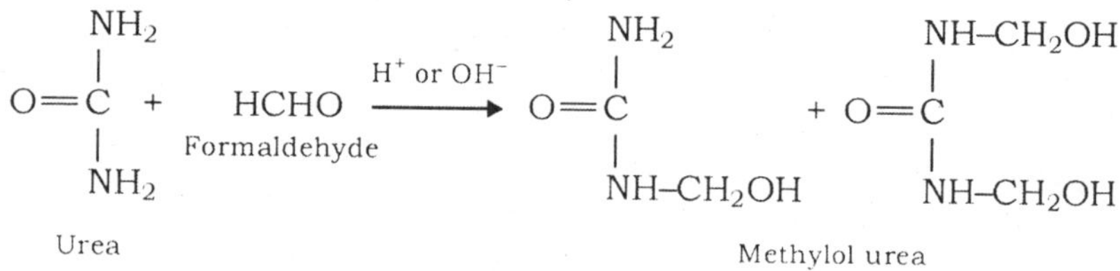

The methylol urea undergoes polycondensation reaction with themselves or with urea to form a cross-linked network polymer similar to the phenol-formaldehyde resin.

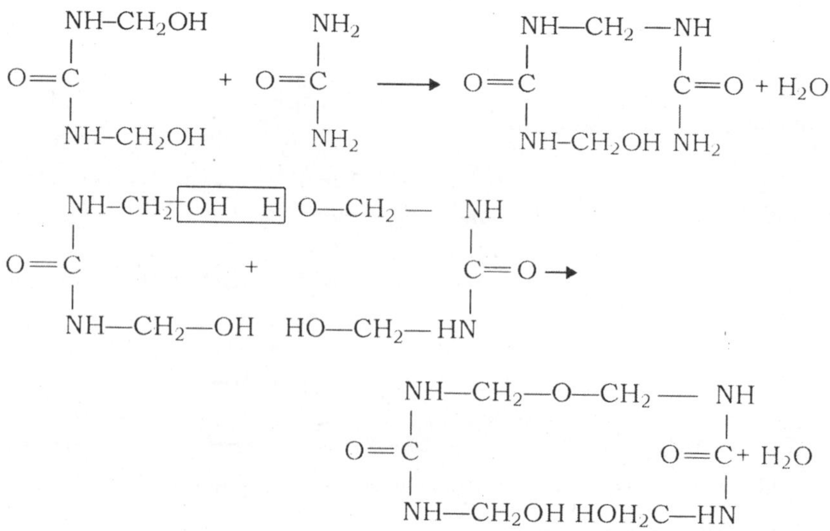

Chemicals: Urea, formaldehyde (40% aqueous solution) and sulphuric acid.

Apparatus: 250 ml beaker or conical flask, hot plate cum magnetic stirrer and thermometer.

Procedure: 10 g of urea and 12.5 ml of formaldehyde solution (40%) is taken in a 250 ml beaker, stirred-well using magnetic stirrer/glass rod until a homogeneous solution is obtained. Add 3-4 drops of dilute hydrochloric acid with constant stirring and reflux for 1–2 hours then add 50 ml of water. The urea-formaldehyde (UF) resin is formed as white powder.

Report: Report the yield, colour and confirmation tests for urea-formaldehyde.

Experiment 13

Preparation of Aniline-Formaldehyde by Polycondensation Method

Principle: Aniline and formaldehyde react under various conditions. In the presence of excess formaldehyde, methylene bridges may be formed to give slightly cross-linked aniline-formaldehyde resin.

Reaction:

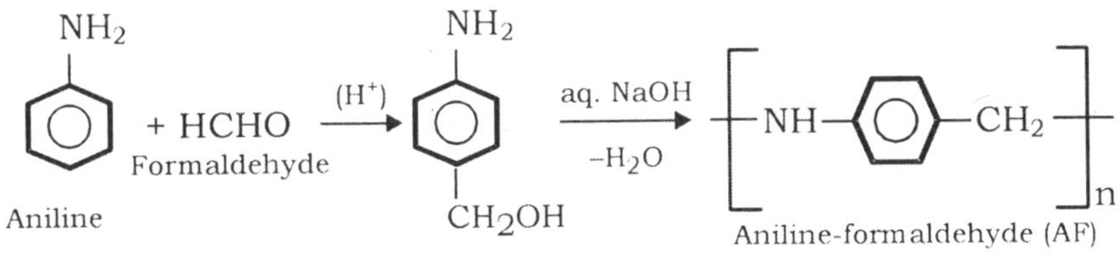

In the presence of excess of formaldehyde, methylene bridges may be formed to give slightly cross-linked products.

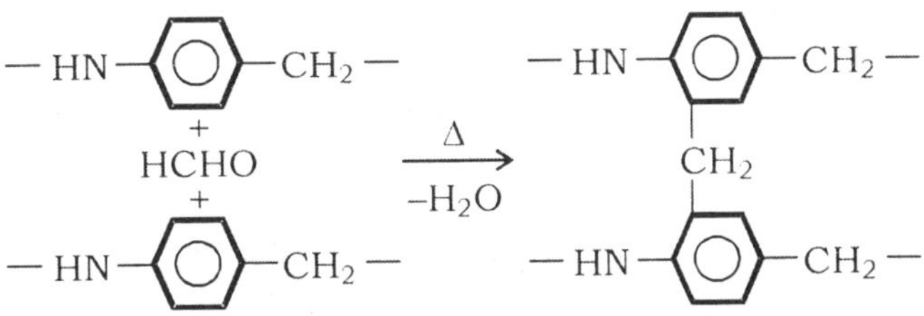

(Cross-linked aniline-formaldehyde)

Chemicals: Aniline, formaldehyde (40 % aqueous solution), dil HCl and 1% NaOH solution.

Apparatus: 250 ml beaker, hot plate-cum-magnetic stirrer and thermometer.

Procedure: 9.3 g of aniline in dil HCl is taken in a 250 ml beaker. 10 ml of formaldehyde solution (40%) is added slowly with stirring. The mixture is then stirred-well for 15 minutes to form an orange-red solution. Then this solution is fed into aqueous sodium hydroxide solution (1%) and refluxed for 30–45 minutes. The aniline-formaldehyde will precipitate out. Filter the resin, wash with water and dry it.

Report: Report the yield, colour and confirmation tests for the aniline-formaldehyde resin.

Experiment 14

Preparation of DGEBA Epoxy Resin using Bisphenol–A and Epichlorohydrine

Introduction: Epoxy can be defined as an oxygen atom that is bridged between two carbon atoms, which are already joined by covalent bond. Epoxy resin available commercially as a monomer. It can be cross-linked using different kinds of hardeners such as diamine, diacid, di-isocyanate, etc. After cross-linking, epoxy resin becomes a thermoset. Epoxy resins are extensively used in structural beams, sport goods, electrical and electronic industries, automobile, aerospace and hydrospace vehicles, paint, coating, military, adhesive, fire retardant, etc. Some of the commercially available epoxies are: DGEBA—diglycidyl ether of bisphenol–A, TGDDM—tetraglycidyl diaminodiphenylmethane, TGPAP—triglycidyl p-aminophenol, etc.

Principle: Epoxy resins are commonly prepared by the reaction between diamine or diol with base induced epichlorohydrine to give an intermediate low molecular weight, essentially linear polymer, having terminal epoxide and pendant hydroxyl groups. The following reactions are believed to occur in the one-step operation leading to fusible pre-polymer.

Reactions:

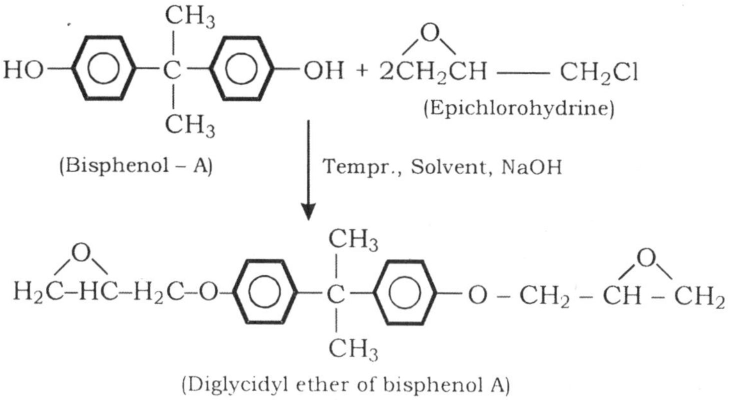

Chemicals: Bisphenol–A, epichlorohydrine, sodium hydroxide and benzene.

Apparatus: Hot plate-cum-magnetic stirrer, dropping funnel, thermometer, three-necked round-bottom flask (resin kettle), vacuum pump and water-cooled condenser.

Procedure: Resin kettle is equipped with stirrer, thermometer, condenser and dropping funnel. A mixture of 22.8 g (1 mole) of bisphenol–A, 22.5 g (10 moles) of epichlorohydrine, 5 ml of water and 5 ml methanol is taken in resin kettle. Temparature of the reaction maintained at 85° ± 2°C for 5-6 hours. A total of 8.2 g (2.5 moles) sodium hydroxide in water (5% solution) is added drop wise slowly through separating funnel after 5 hours of reaction. The heating is stopped when the temperature reaches 80°C and the temperature is controlled so that it does not exceed 100°C. When the reaction temperature falls to 95°C, another 0.5 g of sodium hydroxide is added. Temperature control is exercised as above. The remainder of the sodium hydroxide is added in 0.5 g increments. After the final addition of base, no cooling is applied. When the exothermic reaction subsides, the excess epichlorohydrine is distilled at a pressure of about 50 mm of Hg with a pot temperature not exceeding 150°C. The residue is then cooled to about 70°C and 50 ml benzene is added to precipitate the salt present. The salt is removed by vacuum filtration and washed with 50 ml benzene. The benzene and the excess epichorohydrine is vaccum distilled. When the pot temperature reaches

125°C, a vacuum of about 25 mm of Hg is applied and the distillation continued. The resulting clear and highly viscous liquid is epoxy resin. Calculate the yield, epoxy equivalent weight (EEW), percent hydrolyzable chlorine content, viscosity, specific gravity, colour, etc., of the prepared DGEBA epoxy resin.

Report: Report the yield and confirmation tests for the DGEBA epoxy resin.

Experiment 15

Preparation of Diglycidyl Aniline Epoxy Resin

Principle: Diglycidyl aniline (DGA) is prepared by the condensation of aniline with excess epichlorohydrine in presence of azeotropic mixture of solvents. The following reaction takes place:

Reaction:

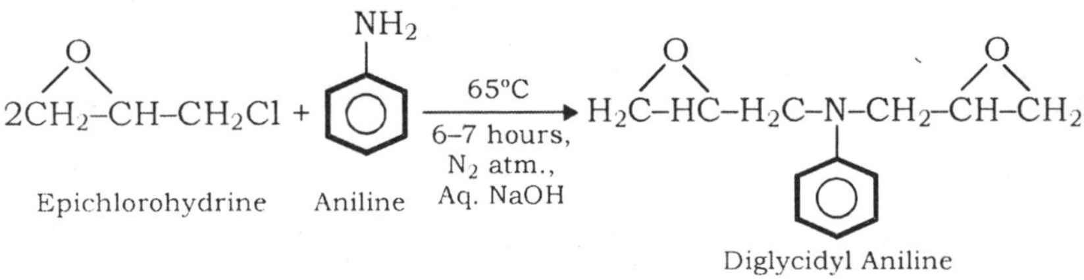

Chemicals: Aniline, epichlorohydrine, methanol, benzene and sodium hydroxide, etc.

Apparatus: 500 ml three-necked round-bottom flask, hot plate-cum-magnetic stirrer, nitrogen gas cylinder, separating funnel, vacuum pump, vacuum oven and vacuum flash evaporator.

Procedure: 38 ml of aniline is added into a mixture of 268 ml epichlorohydrine, 100 ml of 95% methanol and 16 ml of water in a 500 ml three-necked round-bottom flask with nitrogen gas purge. The temparature of the reaction maintained at 75±2°C for 6–7 hours. About 36 g of NaOH in 120 ml of water is added using separating funnel over a period of 20 minutes, at 60°C. The contents are held for an additional 30 min, at 65°C, then stripped under vacuum at 65°C/30 mm Hg to remove excess epichlorohydrine. Then 60 ml of benzene is added to the residue and this solution is washed four times with cold water and two times with hot water. The water layer is removed by decantation and the organic layer is distilled under vacuum at 70°C. A yield of 70–80 %, viscous DGA resin is obtained.

Report: Report the yield, epoxy equivalent weight (EEW), percent hydrolyzable chlorine content, viscosity, specific gravity, colour and confirmation tests for the prepared diglycidyl aniline epoxy resin.

Experiment 16

Preparation of Polyester Resin using Ethylene Glycol and Maleic Acid by Polycondensation method

Principle: Polyesters find commercial use as fibres, plastics and coatings. They are prepared by polycondensation reaction between bi-functional components such as diacids and diols or hydroxy acids with an elimination of water as by-product. Four useful preparation methods are: direct esterification, trans-esterification, reactions of alcohols with acid chlorides and reactions of alcohols with anhydrides. The reactions can be carried out by melting, solution or interfacial polycondensation.

Reactions:

$$\underset{\text{Ethylene glycol}}{HO{-}(CH_2)_2{-}OH} + \underset{\text{Maleic acid}}{HOOC{-}CH{=}CH{-}COOH}$$

$$\Big\downarrow \; \text{5 hours, nitrogen atm., } 180\text{–}200^{\circ}C$$

$$\underset{\text{Polyester}}{HO{-}CH_2{-}CH_2{-}\!(O{-}CO{-}CH{=}CH{-}CO{-}O{-}CH_2{-}CH_2)_n{-}OH} + H_2O$$

Chemicals: Ethylene glycol, maleic acid, *p*-toluene sulphonic acid and xylene.

Apparatus: 250 ml three-necked round-bottom flask, thermometer, separating funnel, Dean and Stark apparatus and Leibig condenser.

Procedure: 50 ml of xylene, 6.2 g of ethylene glycol, 11.6 g of maleic acid (1:1 mole ratio) and 0.1 g of *p*-toluene sulphonic acid catalyst are taken in a 250 ml three-necked round-bottom flask. Two-three pieces of asbestos are added to the flask and it is placed on a electrical heating mantle with an energy regulator. The central neck of the flask is attached to Dean and Stark apparatus fitted with a Leibig condenser and the side necks are fixed to a gas inlet tube and mercury well for the thermometer.

The Dean and Stark apparatus is filled with xylene up to the graduated portion. A slow stream of carbon dioxide gas is bubbled through the contents of the flask to prevent any oxidative degradation of the polyester formed. The flask is heated gently until its contents become homogeneous. Heating is then adjusted such that the contents of the flask start refluxing gently. As the polyester is formed by the polycondensation reaction, the water elimination is carried out along with xylene vapour in the form of an azeotropic mixture. When the azeotropic vapour mixture condenses at the receiver part of the Dean and Stark apparatus, the xylene and water separate out and only xylene returns to the reaction flask. The reaction is carried out until 18 ml of water is collected, when the polyester formation is complete. Next, Dean and Stark apparatus is removed and the xylene is separated by distillation. The liquid polymer obtained is further purified by washing with methanol and drying in vacuum at 60°C. The product obtained is unsaturated polyester, i.e. a polyester resin containing double bonds in its backbone. With a suitable free radical initiator, styrene can cure this polymer by cross-linking.

Report: Report the yield and confirmation tests for the polyester resin.

Experiment 17

Preparation of Nylon-6 from Caprolactum

Principle: Ring opening polymerization of caprolactum in presence of dibasic initiator yields polycaprolactum (Nylon-6).

Reaction:

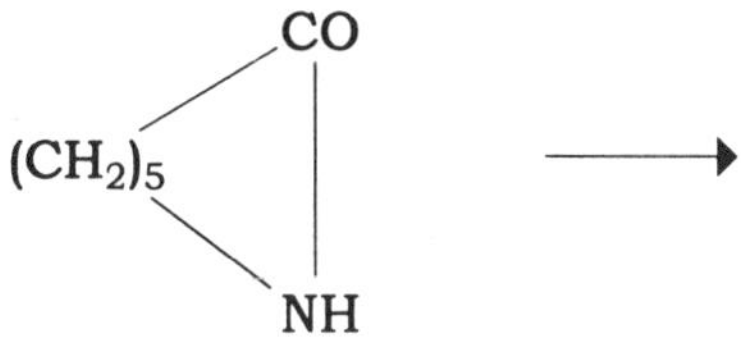

Caprolactum

$$-[HN-(-CH_2)_5-CO]_n-$$

Nylon-6

Chemicals: Caprolactum, adipic acid, methanol, nitrogen (oxygen free).

Apparatus: Aluminum block (9″ × 5″ dia) drilled for reaction tube and thermometer. Pyrex reaction tube (bulb 6 in. × 18 mm ID × 23 mm OD; stem 5 in. × 3 mm ID × 5 mm OD), band heater for block.

Procedure: 11.3 g (0.1 mole) of purified caprolactum (M.P. 68 °C), is weighed into a Pyrex reaction tube together with 0.09 g (0.005 mole) water and 0.15 g (0.001 mole) of adipic acid. By means of a drawn-out capillary tube, a stream of oxygen free nitrogen is applied for 5 minutes before carefully sealing at the stem. The tube is now placed in an aluminum block and held at 250°C for 2 hours. At the end of this time, the tube is cooled and the end is cut off neatly. The tube is attached via rubber tubing to a vacuum pump, replaced in the aluminum block and heated further for an hour at 250°C, then cooled while still under vacuum. A solid translucent plug of polymer should result. It is recovered by breaking the tube. The polymerization only proceeds to an equilibrium conversion of caprolactum, which varies from 90 to 95% depending on the temperature.

Report: The report should include the yield of polymer and melting point of the product.

Experiment 18

Preparation of Nylon-66 from hexamethylene diamine and adipic acid.

Principle: Nylon-66 is prepared from the reaction of diamine and diacid groups, which yield amide groups with the elimination of water. The stoichiometric equivalence required to obtain high molecular weights is insured by the structure and purity of the monomer.

Reaction:

$H_2N—(CH_2)_6—NH_2$	+	$ClOC—(CH_2)_4—COCl$
(Hexamethylene diamine)		(Adiponyl chloride)

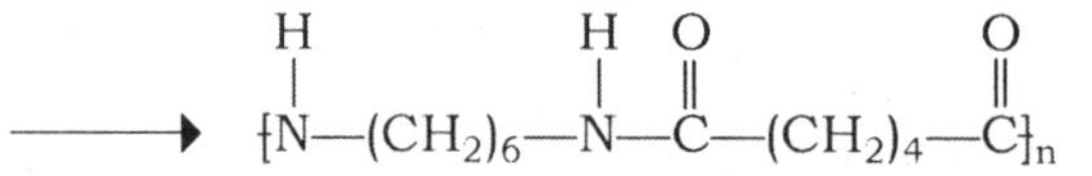

Nylon-66

Apparatus: Aluminium block (9″ × 5″ diameter) drilled for reaction tube and thermometer, Pyrex reaction tube (bulb 6″ × 18 mm I.D. × 23 mm O.D.; stem 5″ × 3 mm I.D. × 5 mm O.D.), band heater for block, separating funnel, Buchner funnel.

Chemicals required: Adipic acid (m.p. 151°C), hexamethylene diamine (m.p. 38°C), 95% ethanol, nitrogen.

Procedure: Preparation of nylon salt (hexamethylene diammonium adipate): Adipic acid (14.4 g, 0.49 mole) is dissolved in 120 ml 95% ethanol and the solution is filtered. Hexamethylene diamine (11.6 g, 0.5 mole) is dissolved in a mixture of 32 ml 95% ethanol and 6 ml distilled water, and the solution is filtered. The adipic acid solution is placed in a 500 ml beaker fitted with a motorized stirrer, and the diamine solution is delivered into the vigorously stirred solution via a separating funnel for15 minutes, stirring being continued for 30 minutes thereafter. The neutralization is exothermic reaction, and the nylon salt should precipitate out as the temperature falls. The white crystalline salt is washed twice by decantation with 95% ethanol, then filtered using Buchner funnel at the pump and air-dried.

This product should be analyzed for melting point (literature value: 183°C) and pH of a 9.5 % solution in water (literature value: 7.62) before proceeding to the next step. If the starting materials are impure, it may be necessary to determine exact equivalency by titration.

Preparation of nylon-66, poly(hexamethylene adipamide)— A mixture of 10 g hexamethylene diammonium adipate and 0.055 g (1 mole %) of adipic acid is weighed into the Pyrex reaction tube. The tube is attached to a vacuum line and outgassed for 5–10 minutes, then purged with oxygen-free nitrogen gas and re-evacuated, and the stem is sealed under vacuum (this operation must be done carefully or the tube is liable to leak or explode). The tube is now placed in the heated aluminium block for 2 hours at 220°C. After this time, the tube is cooled, the neck is cut-off neatly and the tube is connected via rubber tubing to a vacuum pump. The tube is restored to the aluminium block, and the temperature is raised to 270°C and held for 2 hours to complete the polymerization. The tube is removed from the block and cooled while still under vacuum. A translucent plug of nylon-66 should result, from which fibres can be drawn by plucking with the hot wire. The melting point should be determined using the capillary method.

Report: Report the yield and melting point of the nylon salt and nylon-66.

Experiment 19

Preparation of Cellulose Secondary Acetate from Cellulose or Bleached Pulp

Introduction: Cellulose plastics are derived from naturally occuring material. The cellulose can be chemically modified by nitration, acetylation, etc. A variety of chemically modified derivatives of cellulose products are available commercially. The derivatives of cellulose materials find applications in injection moulded articles, films and sheets.

Principle: Fully acetylated bleached pulp is cellulose triacetate, i.e. all the hydrogen of the three hydroxyl groups (functional groups) of cellulose is replaced by acetyl group, (—$COCH_3$) of acetic anhydride. This triacetate is hydrolyzed by the addition of calculated quantity of water along with acetic acid. This will produce cellulose secondary acetate (CSA). This contains 55 ± 1 % by weight of acetic acid and around 2.65 degree of substitution.

Reaction:

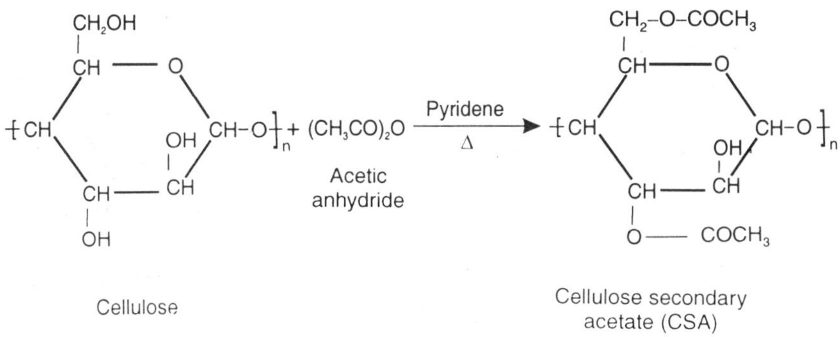

Chemicals: Acetylated cellulose, glacial acetic acid and acetone.

Apparatus: Beaker, thermometer, constant temperature bath and glass rod.

Procedure: About 5g of fully acetylated cellulose is taken in a 100 ml beaker or iodine flask, 18 ml of glacial acetic acid and 5.25 ml of water is added. The above mixture is heated at 50–55°C for one and half hour on a constant temperature bath. The sample is tested for solubility in acetone after one and half hours. If it is fully dissolved in acetone, then the whole mixture is taken out from water bath and water is initially added slowly with constant stirring until the precipitation point, later quickly added with rapid stirring. The precipitate is filtered and washed with water to remove excess of acetic acid and finally dried under vacuum at 60°C for 3 hours.

Report: Report the yield and confirmation test for cellulose secondary acetate.

Experiment 20

Preparation of Cellulose Tri-Acetate from Cellulose

Principle: Cellulose (bleached cotton linter pulp or suitable quality wood pulp) is first pre-treated with glacial acetic acid (100 % pure acetic acid) and then allowed to react with pure acetic anhydride. Sulphuric acid is used as a catalyst. Cellulose reacts with acetic anhydride to give cellulose triacetate. In cellulose triacetate, the degree of substitution is three and combined acetic acid content is 62.5 % by weight.

Reaction:

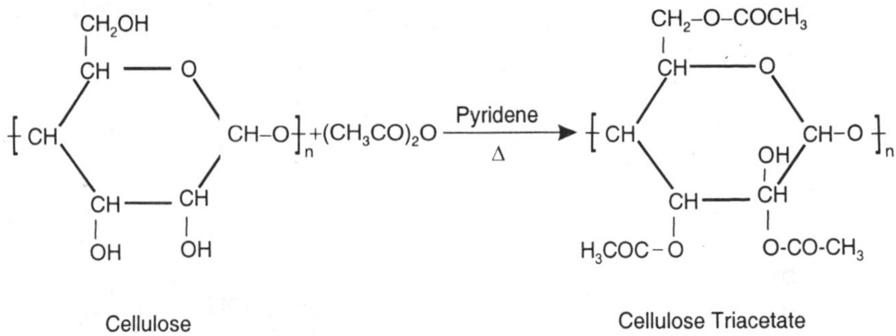

Chemicals: Bleached cotton linter, acetic anhydride, acetic acid and conc. sulphuric acid.

Apparatus: 250 ml iodine flask, beaker, thermometer, constant temperature bath and glass rod.

Procedure: 15 g of cotton linter is loosened and placed in 250 ml iodine flask and add 75 g (71.4 ml) of glacial acetic acid. Then the beaker is placed on water bath for 30 minutes with constant stirring. The beaker is cooled to 30–35°C and then 48 ml acetylating mixture of acetic acid and 2 ml of sulphuric acid are added. Then the mixture is heated and maintained a temperature of 50-55°C for one and half hours with constant stirring until lumps are formed. The water is added slowly to the obtained cellulose triacetate (CTA), with stirring, precipitation of CTA takes place. Then large volume of water is added with stirring. The precipiate of CTA is filtered and washed with water to remove acetic acid. Finally the product is dried in vaccum.

Report: Report the yield and confirmation test for cellulose triacetate.

Experiment 21

Preparation of Chlorosulphonated Polyethylene from Polyethylene

Principle: The chlorosulphonated polyethylene is prepared by the radical displacement reaction on polyethylene.

Reaction:

$$-\!\left(CH_2-CH_2\right)_n- + SO_2 + Cl_2 \xrightarrow[\text{Dark}]{\text{in}}$$

polyethylene

$$\left(CH_2-\underset{\displaystyle Cl}{\underset{|}{CH}}\right)_n-\left(CH_2-\underset{\displaystyle SO_2Cl}{\underset{|}{CH}}\right)_m-\left(CH_2-CH_2\right)_p$$

Chlorosulphonated polyethylene

Chemicals: Low density polyethylene (LDPE), carbon tetrachloride, sulphuryl chloride, liquid chlorine gas cylinder and methanol.

Apparatus: 500 ml three-necked reaction flask equipped with stirrer, reflux condenser, separating funnel and gas inlet tube, Dewar flask and 1 litre beaker.

Procedure: 500 ml three-necked reaction flask attached with a Dewar flask containing solid CO_2 is connected to a chlorine cylinder. Students should read instructions carefully, relating to the use of liquid chlorine cylinder in the laboratory, especially in regard to protecting the cylinder from suck back. Chlorine is passed and condensed until about 50 g is collected. Meanwhile, 25 g of LDPE granules are added into the reaction flask equipped with stirrer, reflux condenser, separating funnel and gas inlet tube. Carbon tetrachloride (500 g) is added and heated under reflux until polyethylene has dissolved. 15 g of sulphuryl chloride is added through the separating funnel, then the chlorine gas is passed from the cold trap through the gas inlet tube. A 200 W-tungsten filament lamp is switched onto illuminate the content of the flask, which is again heated to reflux. After 1 hour, the reaction flask is removed and the contents are poured into a one litre beaker containing 500 ml of methanol. The formed soft tacky precipitate is washed several times with methanol and finally dried in hot air oven at 70°C. The product is weighed and the percentage of yield of elastomer is reported.

Report: Report the yield, chlorine content and confirmation tests for the prepared chlorosulphonated polyethylene.

Experiment 22

Preparation of Sulphonated Polystyrene from Polystyrene

Principle: Sulphonation of polystyrene is an example for an electrophilic substitution reaction on aromatic polyhydrocarbons and can be used to measure the exchange capacity of an ion-exchange resin.

Reaction:

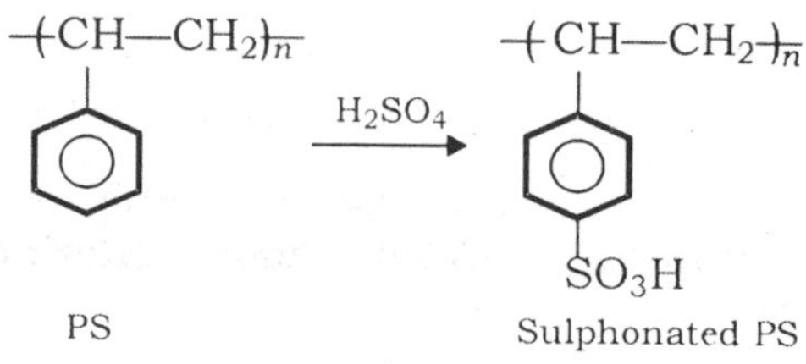

Apparatus: 500 ml three-necked round bottom flask, stirrer, thermometer, separating funnel, water-cooled condenser, one litre conical flask, burette with rubber pinch.

Chemicals: Polystyrene (PS) (cross-linked beads of poly styrene-co-divinyl benzene), sulphuric acid, silver sulphate, NaCl solution (2N), HCl (2N).

Procedure: The 500 ml of three-necked round-bottom reaction flask is assembled with the stirrer, condenser and thermometer. A mixture of 0.2 g silver sulphate and 150 ml concentrated sulphuric acid is heated in the flask at 80–90°C with stirring. Then, 20 g of polystyrene is added. The temperature is raised spontaneously to 100°C and held at this temperature for 3 hours. The reaction flask and contents should be kept undisturbed for 24 hours. Contents of the flask are poured into one litre conical flask containing 50 % sulphuric acid, which is then cooled and diluted slowly with distilled water. The beads are washed thoroughly on a filter, whereupon they should be free of sediment and allow free flow of water. The burette with the rubber pinch clip is fitted up by introducing a light wad of glass wool, then pouring in the suspension of the sulphonated beads. Then, 100 ml of 2 N NaCl is poured through the column over a period of 10 minutes, followed by 100 ml of 2 N HCl, also over a period of 10 minutes. The column is now in the acid form, but free acid may be present, hence the column is washed with distilled water.

Report: Report the yield as the HCl effluent collected as 10 ml aliquots and each aliquot is titrated against N/10 NaOH, using phenolphthalein indicator until the normality of the effluent is reduced to 0.001 N.

Experiment 23

Sheet Casting Using Methyl Methacrylate

Aim: To illustrate the techniques of controlling exotherm and shrinkage during bulk polymerization.

Apparatus: Plate glass sheets of 25 cm × 25 cm × 0.65 cm, cellulose or nylon tubing (1.8 cm lay flat width), spring clips.

Chemicals: Methyl methacrylate, benzoyl peroxide (BPO) or AIBN.

Procedure: Methyl methacrylate is freed from inhibitor by shaking approximately 250 ml of monomer in a separating funnel with 25 ml of an aqueous solution containing 2 g NaOH and 10 g NaCl per 100 ml. The aqueous layer is allowed to separate and run off and the procedure is repeated until the aqueous layer is colourless. The monomer is then washed with successive 10 ml portions of water till this gives no alkaline reaction. The monomer is now allowed to stand over the granular calcium chloride till free of turbidity and is then filtered onto a clean 250 ml conical flask, which should be filled nearly to the top. 0.2 % solution of benzoyl peroxide or AIBN is added to the purified monomer. The flask is lightly corked and held in water bath at 80–90°C for 30 minutes, until the reaction mixture becomes viscous. At this point, the flask should be removed and allowed to stand and thicken while cooling. If the flask remains hot through exotherm or if bubbling commences, it should be cooled under running water. Transfer to another flask is also effective in controlling exotherm, due to inhibition by contact with air. Control of this stage is gained by experience.

Meanwhile, a glass cell is prepared. Two sheets of glass plate are cleaned by rubbing with detergent until the surface is bright and readily wetted by water. The sheets are rinsed thoroughly with tap water and finally with distilled water and placed on edge in the oven at 40°C to dry.

Cellulose or nylon layfalt tube is opened and pulled over a 100 cm of 6 mm diameter rubber tubing, using a few drops of alcohol as a lubricant. This serves as a gasket and is sandwiched between the plates around all four edges using the spring clips to retain it in position. An 8 cm length of gasket is lifted from the upper edge of the cell, and the viscous monomer is carefully poured into the cell through this opening till the cell is nearly full. The length of the gasket is now pushed back into the inclined cell, when the other end of the gasket is pulled down sharply to prevent reentry of air. The cell is then placed vertically on edge in an oven at 40°C or in water bath at 60°C and left there till the next period. When cool, the sheet of PMMA should strip easily on removing the spring clips.

The residue of the syrup should be poured into a test tube, which is corked and placed in an oven or water bath alongside the cell. On complete polymerization, a rod with an irregular surface will have resulted, in contrast to the smooth sheet obtained from the cell.

Report: The report should include the weight, dimensions, and density of the cast PMMA. From the density, the mean thickness may be calculated. This value should be compared with micrometer readings of thickness at various points in the sheet.

Experiment 24

Preparation of an Emulsion Paint Using Vinyl Acetate

Aim: To conduct a "high temperature" emulsion polymerization and to examine a pigmented surface coating.

Apparatus: Four-necked reaction flask, condenser, stirrer, separating funnel, thermometer and nitrogen inlet, ball-mill (1 litre), scratch test apparatus, paint can, No. 0 emery cloth, paint brush, etc.

Chemicals: Vinylacetate, partially hydrolyzed polyvinyl acetate, surface active agent, ammonium persulphate, sodium metabisulphate, carboxymethyl cellulose, sodium salt, sodium polymetaphosphate, titanium dioxide, china clay, dibutyl phthalate, carbon tetrachloride.

Procedure:

Emulsion polymerization: The suspending agent, polyvinyl acetate (5.4 g) and aerosol OT (0.72 g) are dissolved in 100 g distilled water, while 2.86 g of surface active agent is dissolved in 100 g freshly-distilled vinylacetate (BP 72–73°C/ 760 mm Hg). The aqueous solution is placed in a 700 ml four-necked reaction flask equipped with separating funnel, thermometer, nitrogen inlet tube, stirrer and reflux condenser. Nitrogen gas is purged vigorously for about 5 minutes, then reduced to a slow stream that is maintained throughout. The temperature is raised to 65°C by means of a heating mantle and with vigorous stirring, 10 g of the vinylacetate monomer is added via the funnel. An emulsion is formed, and 1 ml of 2 % aqueous solution of sodium metabisulphate is then added. The reaction mixture is maintained at 68–75°C to gently reflux the vinyl acetate, and after 5 minutes, 4 ml of 2% aqueous solution of ammonium persulphate is added through the separating funnel.

After a short induction period, the polymerization should commence and the reflux rate should fall off as the amount of residual monomer decreases. Successive quantities of 20 g vinylacetate monomer is added slowly via the separating funnel whenever the refluxing falls down, until the whole 100 g have been added. If refluxing doesn't subside, a further 2 ml of 2% aqueous ammonium persulphate may be added. Finally, the temperature is raised to 80°C for 10 minutes to complete the polymerization. The reaction should be complete in about 3 hours.

A weighed portion (10 g) of the cooled emulsion may be poured upon a galss plate and left to dry in air to enable the cast film to be weighed and examined.

Formulations of an emulsion paint: Solutions of pigment dispersing agents are first prepared by dissolving 4 g carboxy methyl cellulose, sodium salt in 100 ml of water and by dissolving 0.64 g sodium polymetaphosphate in 16 ml of water. After ensuring that the jar of a 1 litre ball-mill is half-full of graded porcelain balls and is free of contamination, the above solutions are added followed by 260 g of TiO_2 and 150 g china clay. The lid of the ball mill is sealed and the jar is rotated for 24 hours.

At the next period, the ball-mill is fully drained into a 1 pint paint can and a mixture of 190 g polyvinyl acetate emulsion, 30 g dibutyl phthalate and 200 g of water is stirred thoroughly by hand. The paint now should have the correct consistency, but may be thinned with 50 ml water before use.

Scratch resistance of paint film: A panel of tin plate measuring 5″ × 2″ is wiped free of any oil/grease, burnished uniformly with No. 0 emery cloth, finishing with a circular motion, then swabbed with a linen rag (or lint-free paper tissue or cotton wool), soaked in carbon tetrachloride, and allowed to dry. The paint film is applied by brush, allowing atleast one hour to elapse between coats. After applying the third coat, the panel should be allowed to stand till the next period.

The painted panel is clamped to the horizontal sliding panel of the scratch resistance apparatus. The needle should have a hemispherical point of 1 mm diameter and the needle should be adjusted normal to the point film. After the paint film has been set into motion at the rate of 3–4 cm/sec, the needle is brought gently into contact with the moving panel for a distance of 6 cm. The needle is then raised and the paint film examined. The pan is loaded with progressively increasing weights and the procedure is repeated until scratch marks are noted.

Report: The report should include the yield of polyvinyl acetate and the solid content of the emulsion. The smoothness and brushability of the paint should be examined. The hardness of the derived paint film should be expressed in terms of the minimum load required to cause scratch marks.

Experiment 25

Preparation of Alkyd Resins

Aim: To prepare the alkyd resin from phthalic anhydride.

Apparatus: Two pyrex boiling tubes, 360°C thermometer, Bunsen burner.

Chemicals: Phthalic anhydride, glycerol, ethylene glycol.

Procedure: The following mixtures are heated in separate tubes for 90 minutes at 160–180°C (the experiment should be carried out in a fuming cupboard): (a) 4.8 g (1.5 mol) phthalic anhydride + 6.1 g (5.6 ml) (1mol) glycerol; (b) 14.8 g (1 mol) phthalic anhydride + 6.2 g (5.6 ml) (1 mol) ethylene glycol.

After this treatment, a drop of each resin should yield a brittle bead on cooling with water. Approximately half of each resin is cooled and retained for solubility tests.

The remainder of the mixtures are heated to 240°C and maintained at this temperature under observation. The glycerol resin gels indicating thermosetting properties. While the ethylene glycol resin remains mobile.

Small amount of the samples should be tested for hot and cold solubility in acetone, ethyl alcohol, linseed oil and white spirit and the results recorded.

Alkyd resins are used in plastic moulding compositions on account of their good surface wetting properties. The oil modified types are extensively used in enamel varnishes and paints.

Report: Report the colour and yield of the alkyd resin.

Experiment 26

Preparation of Polyvinyl alcohol from Polyvinyl acetate (Polymer modification)

Aim: To prepare polyvinyl alcohol from polyvinyl acetate by using the polymer modification technique.

Apparatus: 500 ml round-bottom flask, reflux condenser, $CaCl_2$ guard tube, one litre beaker and mechanical stirrer.

Reaction:

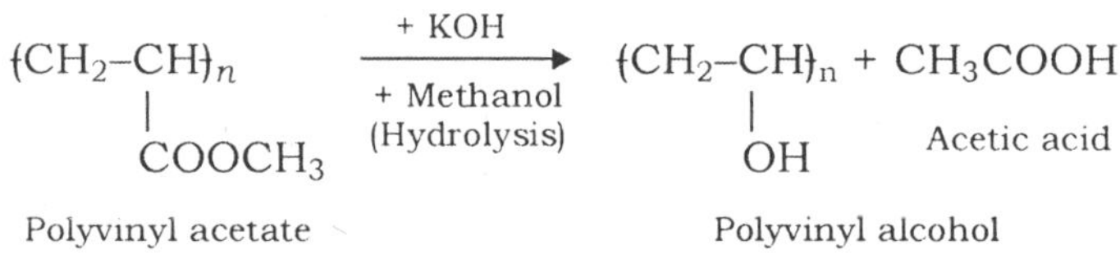

Chemicals: Polyvinyl acetate (low viscosity grade), anhydrous methanol, potassium hydroxide and acetone (water-free).

Procedure: Into a 500 ml round-bottom flask fitted with a reflux condenser and protected from moisture by an anhydrous calcium chloride guard tube are placed polyvinyl acetate (8.6 g, 0.1 mol) and anhydrous methanol (200 ml). The contents are heated on a water bath for about 20 minutes to dissolve polyvinyl acetate and then potassium hydroxide (2.5 g) is added and the refluxing is continued for another one hour. Polyvinyl alcohol (PVA) separates out as a fine powder.

To purify polyvinyl alcohol (PVA), it is first filtered using the Buchner filter pump and an alkali resistant filter paper. It is finally washed several times with anhydrous methanol. It is dissolved in 150 ml of distilled water by heating at 80°C. The solution is then poured drop wise into vigorously stirred anhydrous acetone (500 ml). The precipitated polyvinyl alcohol is then filtered using the pump and dried in a desiccator.

Report: The report should include yield, and confirmation tests of prepared PVA.

Experiment 27

Depolymerization of Polymethyl methacrylate

Aim: Depolymerization of PMMA and to get back the monomer.

Apparatus: Vacuum distillation assembly, two traps, Dewar flasks, sand bath or a metal alloy bath, thermometer (0–450°C)

Chemicals: PMMA, solid CO_2 and acetone.

Procedure: A vacuum distillation assembly is setup and in distilling flask 20–30 g of PMMA is placed. The receiver is attached to a vacuum system of 0.5 to 2 mm of Hg capacity through a series of two traps cooled with dry ice-acetone cooling mixture or a freezing mixture. The flask is heated by means of a suitable sand or metal alloy bath to 300–400°C and then monomer vapours are collected under vacuum for a period of 3–4 hours. The combined distillate is purified in a suitable distilling apparatus and identified them by determination of boiling point, refractive index and specific gravity.

Report: The report should include the yield of the distillate and characteristic properties of obtained monomer.

EXCERCISE

1. Distinguish between:
   (*a*) Thermoplastic, thermoset and elastomer.
   (*b*) Macromolecules and low molecular weight compound.
   (*c*) Addition and condensation polymerization.
   (*d*) Initiator and inhibitor.
   (*e*) Suspension and emulsion polymerization.
   (*f*) Mass and solution polymerization.
   (*g*) Natural and synthetic polymers.
   (*h*) Homo- and heteropolymers.
   (*i*) Novolac and resol resins.
   (*j*) HIPS and PS.
   (*k*) Amorphous and crystalline polymers.
2. How polymers are classified?
3. Mention the advantages and disadvantages of the polymerization methods like bulk, emulsion, suspension, solution, plasma and solid-gas phase polymerization.
4. Give three to five examples for the following: Thermoplastic, thermoset, elastomer, fibre, initiator, emulsifier, inhibitor,

chain transforming agent, suspending agent, copolymers, free radical initiator, redox initiator, ionic initiator, condensation polymer, Zeigler-Natta catalyst, etc.

5. Name the starting material to prepare the following polymers:
   Nylon-6, Nylon-6,6, Nylon-6,10, epoxy (DGEBA and TGDDM), PET, PBT, SAN, PF, PC, PPS, PEEK, PEN, PPO, etc.
6. Define the term epoxy. Give some examples. Mention their salient features and commercial uses.
7. Define the term copolymer. Distinguish between homo polymers and copolymers.
8. Name different types of copolymer. Give examples.
9. Name the different methods of polymerization.
10. Mention the different steps in the polymerization.
11. How do you initiate the polymerization process? Based on the initiator how do you classify the polymerization reaction?
12. Name the different ways of producing free radical of the initiator. Explain.
13. Explain the different ways of termination of the polymerization process.
14. Name the initiators used for the polymerization methods like, bulk, solution, suspension and emulsion.
15. What is ring-opening polymerization? Explain with an example.
16. List the necessary ingredients used in emulsion polymerization recipe. Explain their functions.
17. Define and describe micelles and discuss their role in emulsion polymerization.
18. What is tacticity? Explain the classification of the polymers based on tacticity, with example.
19. Explain the term dead polymer with an example.
20. Explain the term living polymer with an example.
21. Explain the mechanism of free radical polymerization.
22. Define the concepts of polymer, monomer and functionality. Explain with examples.
23. What is the functionality of the monomer? What is the influence of functionality on the types of products?
24. Mention the functionality of the following:
    aniline, phenol, acetylene, butadiene, urea, formaldehyde, bisphenol-A, DGEBA, propylene, acrylonitrile, etc.

25. The product obtained in bulk polymerization is generally pure. Why?
26. Ring structure containing three atoms is less stable. Why?
27. Why heat dissipation in solid phase polymerization is very poor?
28. Why thermosetting polymers cannot be recycled?
29. What is interfacial polymerization? Explain with examples.
30. Write the repeating units and one commercial example for each of the following:
    (*a*) Polyamide (*b*) polyester (*c*) polyethers
    (*d*) polyanhydrides (*e*) epoxy (*f*) polyurethane (*g*) polyimide
31. Pearl- or bead-shaped products is obtained in suspension polymerization? Give reasons.
32. Which of the following polymers are addition and condensation polymers? Why?
    (*i*) PU (*ii*) PS (*iii*) ABS (*iv*) PVC (*v*) Nylons (*vi*) PVC
    (*vii*) PET (*viii*) SAN (*ix*) EPDM (*x*) PC (*xi*) PPO (*xii*) NBR
33. Is it possible to get polymer in a powdery form from emulsion polymerization? If yes, give reason.
34. Account for the following:
    (*i*) High molecular weight by emulsion polymerization.
    (*ii*) High purity by mass polymerization.
    (*iii*) Living polymers are obtained from ionic polymerization.
    (*iv*) Broad molecular weight distribution (MWD) by bulk polymerization.
    (*v*) low molecular weight and narrow MWD by suspension polymerization.
35. How many types of PP are available commercially? Give their structures.
36. What is the significance of termination process?
37. What is co-ordination polymerization? Name few catalysts are used in this process. Explain the significance of co-ordination polymerization.
38. Define the term thermoplastic elastomer (TPE). Mention their advantages.
39. Why silicone rubbers are highly heat-resistant polymers?
40. Explain the factors affecting on the polymerization process.
41. Name the different kinds of polyolefins available commercially.
42. Distinguish between LDPE and HDPE.

43. Carbon tetrachloride is a very strong chain transfer agent for styrene polymerization. Explain.
44. Distinguish between the terms stereo-regular, stereo-specific and stereo-selective.
45. What types of intermolecular forces are present in polymers? Illustrate.
46. Distinguish between configuration and conformation.
47. Mention the properties and uses of acrylates.
48. Name the uses of solution and emulsion polymerization products.
49. What is precipitation polymerization? Explain.
50. What are the advantages of plasma polymerization?
51. How do crystallinity and molecular weight control the properties of PE?
52. Compare the molecular structural features of LDPE, HDPE and LLDPE.
53. Why is LLDPE preferred over LDPE for various applications?
54. What are the functions of the three components, styrene, acrylonitrile and butadiene, in ABS resins.
55. What are the important performance characteristics of PC? Justify its use in safety helmets and appliances.

3 Analysis and Characterisation of Polymers

INTRODUCTION

Qualitative and quantitative estimations provide enormous support to the plastic industry in quality control, development of new materials, product designing and process optimization. A number of analytical and characterization techniques are available to evaluate polymer characteristic properties. The measurement of density can be simple but it has great importance in quality control and screening of material for specific applications. The size of polymer particles and their distribution when available in powder form can greatly affect the processability and product quality.

Additives are used to achieve specific combinations of properties of polymeric compounds. Their performance depends upon the quality, quantity and required dispersion in the polymer matrix. It is, therefore, important to analyse the additives so that their effective use in the material is ensured. Inherent properties like molecular weight, molecular weight distribution, thermal and rheological behaviour are the prime factors of polymer quality. Data on them are essential for development of new materials, selection of processing conditions, prediction of product performance and quality control.

Various standard methods are available for analysis and characterization of polymers. This chapter deals with some of the important but simple analytical methods as well as some sophisticated techniques for characterization of polymeric materials and compounding ingredients.

EXPERIMENTS

Experiment 1
Estimation of Some Physical Properties of Polymers

(a) Determination of percentage of filler/fibre content in the polymer composite

Principle: Fillers are frequently used for modifying the properties of resins and also used for reducing the cost of the product. Fillers can be organic or inorganic materials. The most important inorganic fillers include quartz, slate, flour, asbestos, chalk, talc, mica and glass. Thcy occur in the form of particulate fillers, flakes, fabric, fibre, wool, mats and ropes. Organic fillers include wood flour, cellulose in the form of fibres, paper shreds and widths, also textile shreds and widths, and synthetic fibres of all types.

Inorganic filler content is determined by ashing and is most readily identified by analysis of the residue. Organic fillers are isolated by solvent extraction (by Soxhlet extraction) method from the resin. The type of fillers can be identified by subsequent analysis.

Apparatus: Silica crucible, electronic balance, muffle furnace and desiccator.

Materials: Composite specimens.

Procedure: The empty crucible is cleaned well and dried. The crucible is weighed in an analytical balance with an accuracy of 0.0001 g (W_1 g). The crucible is filled with a small quantity ($\cong$ 5g) of the supplied composite sample. Once again its weight is recorded (W_2 g). Now, the crucible is kept in the furnace maintained at 600–700°C for 4–5 hours. After the completion of pyrolysis process, the crucible is cooled well with desiccator and the weight (W_3 g) is recorded. The percentage of filler/fibre content of the polymer composite can be calculated using the following equation:

$$\text{Percentage of filler content} = \frac{(W_3 - W_1)}{(W_2 - W_1)} \times 100$$

Result: Filler content of the composite = __________%.

(b) Determination of ash content of the given polymer

Apparatus: Silica crucible, electronic balance, muffle furnace and desiccator.

Materials: Vulcanized rubber.

Procedure: The empty crucible is cleaned well and dried. The crucible is weighed in an analytical balance with an accuracy of 0.0001 g (W_1 g). The crucible is filled with a small quantity of the vulcanized rubber. Once again its weight is recorded (W_2). Now, the crucible is kept in the furnace maintained at 600-800°C for 4–5 hours. After the completion of pyrolysis of the rubber, the crucible is cooled well with desiccator and the weight (W_3) is recorded. The ash content of the rubber is due to the inorganic filler in the rubber sample. The percentage of ash content of the rubber sample can be calculated using the following equation:

Percentage of ash content = $(W_3 - W_1)/(W_2 - W_1) \times 100$

where, $W_3 - W_1$ is the wt. of residue and $W_2 - W_1$ is the weight of the sample taken.

Result: Ash content of the polymer sample = ...%.

(c) Determination of moisture content of the polymer sample

Introduction: The hygroscopic nature of plastic materials causes processing as well as dimensional stability problems. Moisture content characteristics of plastics depend largely upon the basic type and final composition of a material.

Significance: The test serves to setup a suitable drying schedule for hygroscopic polymers to avoid problems during processing. Even a small amount of water can generate large amounts of steam at the high temperature encountered during processing. This steam becomes trapped in the melt, and will expand on decomposition or decompression of melt in later stages leading to voids in the finished product. Such voids are sometimes flattened out through shear during flow, leading to "mica marks". Different test methods are used for quantitative and qualitative determination of moisture in plastic. The gravimetric method is simple to use for quantitative determination of moisture content.

Apparatus: Electronic balance, petridish, desiccator and hot air oven.

Materials: Polymer sample.

Procedure: The test is carried out as per IS-7328 specification. Petridish is cleaned well and dried. Its empty weight is noted down (W_1 g). Approximately 10–15 g of the polymer sample is taken in the petridish. Once again its weight is recorded (W_2 g). The petridish is introduced into the hot air oven maintained at 100 ± 1°C and kept at the same temperature till constant weight is reached in consecutive weighing. Once again the weight of the petridish is noted (W_3 g). The percentage of moisture content of the polymer sample can be calculated using the following equation:

Percentage of moisture content = $(W_2 - W_3)/\ (W_2 - W_1) \times 100$

where, $(W_2 - W_3)$ is the weight loss and $(W_2 - W_1)$ is the weight of the sample taken.

Result: Moisture content of the polymer sample = ...%.

(d) Determination of water absorption of the polymer sample (ASTM D-570)

Introduction: The tendency of plastics to absorb moisture may be overlooked but even the slightest amount of water can significantly alter some key mechanical, electrical or optical property.

Significance: The test serves the purpose of choosing the polymer or to modify the polymer as per the requirement and the working conditions. Depending upon the type and final composition of the polymer even a small amount of water can alter the key properties to undesirable values. Water absorption characteristics of plastic materials are altered by the addition of additives such as fillers, glass fibres, and plasticizers. Some plastics absorb very little water at room temperature, but at higher temperatures, they tend to take up a considerable amount of water and lose properties rapidly.

Apparatus: Electronic balance and hot air oven.

Materials: Polymer sample.

Test Specimen: The test specimen for moulded plastics shall be in the form of a disk 50.8 mm in diameter and 3.2 mm thick. Permissible variations in thickness is ± 0.18 mm for hot-moulded and ± 0.30 mm for cold-moulded or cast materials. The test specimen for sheets shall be in the form of a bar 76.2 mm long by 25.4 mm wide and by the thickness of the materials. The test specimen for rods shall be 25.4 mm long, and 12.7 mm long for larger-diameter rods. The diameter of the specimen shall be the diameter of the finished rod. The test specimen for tubes less than 76 mm in inside diameter shall be the full section of the tube and 25.4 mm long. For tubes 76 mm or more in inside diameter, a rectangular specimen shall be cut 76 mm in length in the circumferential direction of the tube and 25.4 mm in width lengthwise of the tube.

Conditioning: Specimens of materials whose water-absorption value would be appreciably affected by temperatures in the neighbourhood of 11°C, shall be dried in an oven for 24 hours at 50 ± 3°C, cooled in a desiccator, and immediately weighed to the nearest 0.001 g. Specimens of phenolic products and similar products whose water-absorption value not to be appreciably affected by temperatures upto 110°C shall be dried in an oven for 1 hour at 105–110°C.

Procedure:

(*i*) *24-hour immersion:* The conditioned specimens shall be placed in a container of distilled water maintained at a temperature of 23 ± 1°C, and shall rest on edge and be entirely immersed. At the end of 24 ± 1/2 hours, the specimens shall be removed from water once at a time, all surface water wiped off with a dry cloth, and weighed to the nearest 0.001 g immediately. If the specimen is 1.6 mm or less in thickness, it shall be put in a weighing bottle immediately after wiping and weighed in the bottle.

(*ii*) *2-hour boiling water immersion:* The conditioned specimens shall be placed in a container of boiling distilled water, and shall be supported on edge and be entirely immersed. At the end of 120 ± 4 minutes, the specimens shall be removed from water and cooled in distilled water maintained at the room temperature. After 15 ± 1 minutes, the specimens shall be removed from water, once at a time, all water on the surface removed with a dry cloth, and the specimens weighed to the nearest 0.001 g immediately. If the specimen is 1.6 mm or less in thickness, it shall be weighed in a weighing bottle.

Result: The report shall include the procedure of testing and thickness of the material and average value of water absorption.

(e) Determination of melting point of the polymer sample

Apparatus: Melting point apparatus, glass capillaries and thermometer.

Chemicals: Polymer sample.

Procedure: The given polymer is cut into thin section and is inserted in the capillary tube where the other end is sealed properly. The capillary tube is introduced to the melting point apparatus. When the heating element is switched on, the temperature rises gradually. With the help of the eyepiece provided in the apparatus the steady state conditions of the temperature is observed. The temperature at which there is a marked change in phase of the material is noted. The experiment is carried out in duplicate and the average value is recorded as the melting point of the polymer supplied. The melting point of the substance is matched with the literature value.

Result: Melting point of the polymer sample = ...°C

(f) Determination of density of the polymer

Determination of the density of the solid plastics (excluding cellular plastics) is detailed in ISO 1183 (ASTMD 792), which gives four methods according to the nature of the material. Procedure for determination of density by displacement method and gradient techniques are given below:

(*i*) **Displacement Method:** This method is used for specimen in the form of sheet, rod, tube or moulded articles.

Apparatus: Electronic balance, calibrated jar and sinker.

Materials: Polymer sample and distilled water.

This method is suitable for film, sheet, rod, tube and moulded articles. Density is determined at standard laboratory temperatures. A specimen with smooth surfaces free from greases and dust and preferably of mass approximately 50 g is weighed in air (W_1 g) and then in distilled water (W_2 g). If the density of the material under examination is less than that of water, a sinker is attached and the mass of the sinker in water is determined (W_3 g). The mass of the material with sinker is noted (W_4 g). Fine wire is used for the suspension of the specimen.

$$\text{Density} = W_1 / (W_1 - W_2) \times P$$

$$= \frac{\text{Weight of the sample in air} \times P}{\text{Weight in air} - \text{Weight in water}}$$

where, $P = 1000 \text{ kg/m}^3$

If sinker is used:

$$\text{Density} = \frac{W_1 \times P}{W_1 + W_3 + W_4}$$

(*ii*) **Density by Gradient Technique (ASTM D1505):** This test method is developed to determine the density of plastics based on observations of the level to which the specimen sinks in a liquid column exhibiting density gradient in comparison with the standard specimens of known density. A number of calibrated glass beads of precisely known density are introduced into the density gradient. They are allowed to sink in the column to a point where the glass beads float density matches that of the solution. A series of such floats of differing densities within the range of the column serve as a means of calibrating the column. The float position versus float density is plotted on a chart large enough to be read accurately to $\pm$ 1 mm to obtain a calibration line. A density measurement of a specimen can be made by introducing it in the column and measuring the position upon reaching equilibrium. Which, when referred to the calibration line gives an accurate measurement of density. Alternately, density determination requires numerical calculation.

Small granules or cut pieces of polymer can be used for density measurement. A suitable device (e.g. a cathetometer) can be used to measure the position of the specimen in the column.

Table 3.1 lists a number of liquid systems recommended for use in density gradient columns.

Fig. 3.1, two non-reactive miscible liquids are allowed to fill and form a density gradient column in the glass tube. The rate of mixing of liquids and their filling in the tube is suitably controlled. The density gradient column is calibrated with standard glass floats. A commercially available apparatus is shown in Fig. 3.2.

Table 3.1: *Liquid systems recommended for use in density gradient columns*

Solvent System	Density Range (g/cc)
Methanol–benzyl alcohol	0.80 – 0.92
Isopropanol–water	0.79 – 1.00
Isopropanol–diethylene glycol	0.79 – 1.11
Ethanol–carbon tetrachloride	0.79 – 1.59
Ethanol–water	0.79 – 1.00
Toluene–carbon tetrachloride	0.87 – 1.59
Water–sodium bromide	1.00 – 1.41
Water–calcium nitrate	1.00 – 1.60
Zinc chloride–ethanol-water	0.80 – 1.70
Carbon tetrachloride–1,3-dibromopropane	1.60 – 1.99
1,3-dibromopropane–ethylene bromide	1.99 – 2.18
Ethylene bromide–bromoform	2.18 – 2.89
Carbon tetrachloride–bromoform	1.60 – 2.89
Trichloroethylene–bromoform	1.55 – 2.89

Density being sensitive to temperature change, as is the density gradient formed by the liquid mixture, it is important to control the test conditions for density measurements. An entrapped air bubble in the test specimen affect the test results.

(g) Determination of bulk density for polymer powder or granules

Bulk density is defined as the 'weight per unit volume of a material, including inherent voids in the material' as tested. Bulk density is commonly used for materials such as moulding powders.

Apparatus: Electronic balance and specific gravity bottle.

Chemicals: Polymer sample, benzene, acetone and distilled water.

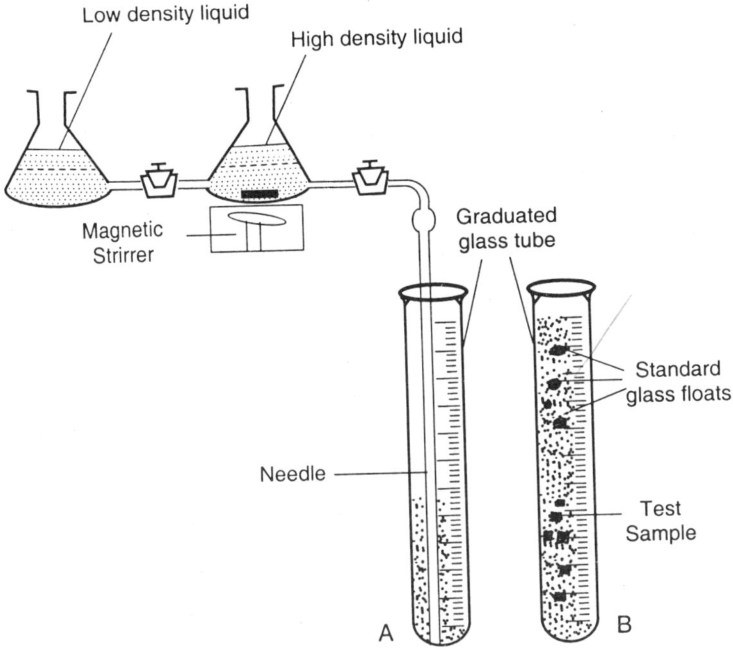

Fig. 3.1: *(A) Setup for the preparation of density gradient column and (B) Calibrated density column*

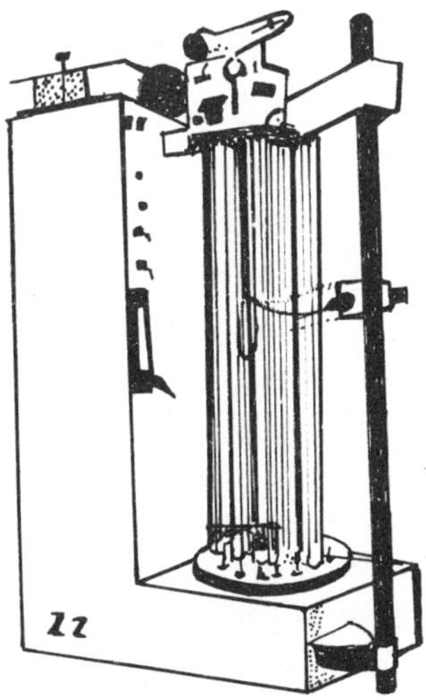

Fig. 3.2: *Density gradient column*

Procedure: The cleaned, dried empty specific gravity bottle is accurately weighed (W_1 g). Then the specific gravity bottle is filled with water and weighed (W_2 g). The specific gravity bottle is filled with benzene and recorded the weight (W_3 g). The weight of the specific gravity bottle with three-fourth volume filled with the given polymer sample and is weighed (W_4 g). The remaining one-fourth volume in the specific gravity bottle is filled with benzene (non-solvent) and is weighed (W_5 g). The specific gravity of polymer powder is calculated by using the following formula:

$$\text{Specific gravity} = \frac{(W_4 - W_1) \times \dfrac{(W_3 - W_1)}{(W_2 - W_1)}}{(W_4 - W_1) + (W_3 - W_5)}$$

where, W_1 = Wt. of empty specific gravity bottle

W_2 = Wt. of specific gravity bottle + water

W_3 = Wt. of specific gravity bottle + benzene

W_4 = Wt. of specific gravity bottle + three-fourth substance

W_5 = Wt. of specific gravity bottle + three-fourth substance + one-fourth benzene

Real weight is defined as the product of real density (specific gravity) and volume of polymer (volume of specific gravity bottle).

Real weight = Volume of specific gravity bottle × Real density

Apparent density is defined as the 'weight per unit volume of a material including voids inherent in the material' as tested. Apparent density is a measure of the fluffiness of the material. This property of plastic moulding powder and granules is useful with respect to their handling in packaging and fabrication. In the case of moulding powder, apparent desnity is termed as bulk density.

ASTMD-1895 specifies three test methods for the determination of apparent density which can be selected according to the size and shape of agglomerates in powder, granule or other form.

The apparent weight of a substance can be measured by filling the polymer powder loosely in to 50 ml measuring jar and record the weight.

$$\text{Apparent Weight} = \left\{\begin{array}{l}\text{wt. of the substance in} \\ \text{50 ml measuring jar}\end{array}\right\} = (W_2 - W_1)$$

where, W_1 = and weight of the empty 50 ml jar; and
W_2 = and weight of the sample + jar

Bulk factor of the material directly affects the perform size, the loading space in the cavities, and the design of the loading trays. Any change in the bulk factor can cause serious moulding problems such as porous or incomplete mouldings, since the moulding charge is measured on volume basis rather than the weight. The bulk factor is defined as the ratio of the volume of any given quantity of the loose plastic material to the volume of the same quantity of the material after moulding or forming. The bulk factor is also equal to the ratio of density after moulding or forming to the apparent density of the material as received.

The value is calculated as follows:

$$\text{Bulk factor} = \frac{\text{Average density of moulded part}}{\begin{array}{c}\text{Average apparent density of} \\ \text{material prior to moulding}\end{array}}$$

Result: Report the bulk factor of the given polymer or filler.

Experiment 2

Determination of Rate of Burning extent and Time of Burning of Flexible Plastic in a Vertical Position.

Significance: The rate of burning of flexible plastic test is measured as per ASTM D-1433. This method covers a small-scale laboratory screening procedure for comparing the relative rate of burning and/or extent and time of burning of plastics in the form of flexible, thin sheets or films, tested in the vertical position. This test method provides a means of comparing the rate of burning and/or extent of burning and time of burning of plastics in the form of flexible thin sheets. Excessive shrinkage of the specimens during the test may invalidate the test results.

Apparatus: Shield (300 mm × 300 mm × 760 mm metal sheet with open top), stand, clamp, burner, timing device.

Test Specimens: 25 mm × 440 mm dimension polymers or film specimens of 10 in numbers are taken. Gauge mark shall be drawn across the specimen 70 mm from one end, defining a 38 cm gauge length over which the burning rate is to be measured.

A typical apparatus for the test is shown in Fig. 3.3. ASTM D-1433 also specifies a test method for flexible plastics in the form of film or thin sheeting which are supported on a 45° inclination. This test method can also be used for rigid and non-rigid sheets, which can be bent through a 45° angle as per the test procedure. Fig. 3.4 shows the test assembly indicating the position of igniter and test specimen.

Procedure: Clamp the specimen vertically so that 43 cm of it is exposed below the clamp, gauge mark at the top and adjust the bunsen burner to provide a flame about 25 mm in total height. Start the timer, apply the tip of the flame to the end of the specimen until it is ignited (t′ sec), but not longer than 15 seconds. The flame may be momentarily withdrawn as needed to establish that the specimen is ignited. If the specimen continues to melt or shrink away from the flame as far as the gauge mark, the test is not applicable. If the specimen ignites, close the shield; stop the timer when any part of the burning edge reaches the gauge mark, or when the flame is extinguished, and note the time (t sec). If the flame is

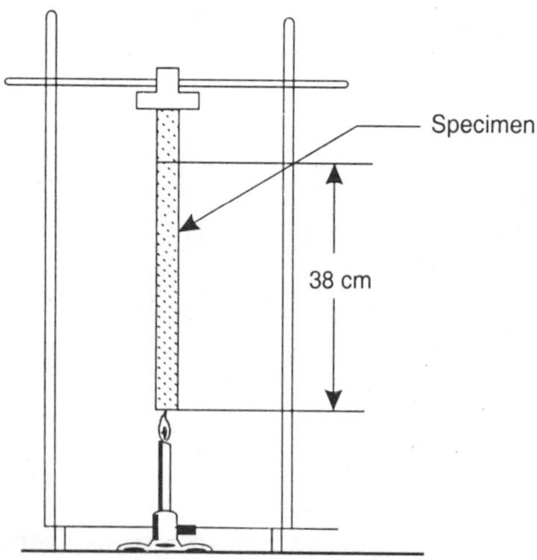

Fig. 3.3: *Apparatus for burning test*

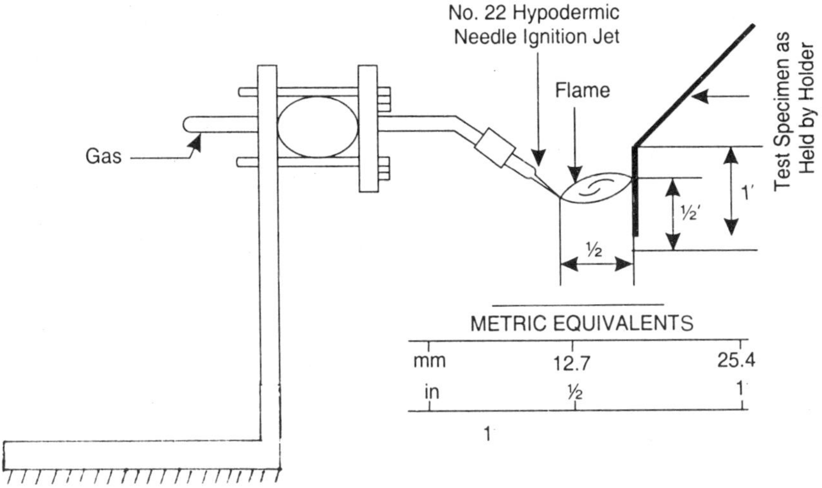

Fig. 3.4: *Igniter assembly*

extinguished before reaching the gauge mark, measure the length (l) in cm from the gauge mark remaining unburned. Note the apparent cause of extinguishment, such as melting and dripping, smothering, etc. Repeat the experiment until three specimens have burned to or beyond gauge mark, or until ten specimens have been tested. If only one of ten burns to the gauge mark, test an additional ten specimens.

Calculations: (i) *Burning rate:* If a specimen burns to the gauge mark, calculate the burning rate in cm per minute by dividing 38 cm by the time in minutes (t/60).

Average time of burning (ATB) and average extent of burning (AEB): If the specimen does not burn to the gauge mark do not calculate the burning rate. But report the mean burning time and length of burning.

Average time of burning (ATB) = $\Sigma(t - t')$ / number of specimens

Average extent of burning (AEB) = $\Sigma(38 - l)$/number of specimens

Material or film	*Guage length 'l' (cm)*	*Time to ignite (s)*	*Time to ignite (min)*	*Burning rate (cm/ min)*

Report: Report the thickness of the material, burning rate or ATB and AEB.

Experiment 3

Determination of viscosity average molecular weight and size of the molecule by viscometric method

Introduction: Any high polymer material is characterized by its usual high molecular weight. Molecular weight of a high polymer is usually expressed as (i) Number average molecular weight ($\bar{M}_n$) which is statistical average of number of monomer units in that polymer or (ii) Weight average molecular weight ($\bar{M}_w$) which is statistical average of molecular weight of different species in the polymer. Colligative properties like elevation of boiling point, depression of freezing point, relative lowering of vapour pressure and osmotic pressure gives number average molecular weight and the light scattering method gives weight average molecular weight. There are other methods like intrinsic viscosity, size exclusion chromatography and vapour phase osmometry are not absolute methods in the sense that they neither give weight average nor number average; but they indicate relative molecular weight. Because of the simplicity and rigidity, intrinsic viscosity of a polymer solution is used as an index of degree of polymerization.

In this background, intrinsic viscosity method is used extensively as a measure of molecular weight and is a relative quality control test for any polymer preparation.

Theory: Viscosity of a polymer solution increases proportionately as the molecular weight of the polymer increases. Thus, viscosity of 0.5% solution of polystyrene (PS) of molecular weight 30,000 will be considerably lower than that of a 0.5% solution of natural rubber (NR) of molecular weight of about 1,40,000.

According to Newton's law of viscous flow, the frictional force F that resists the flow of any two adjacent layers of the liquid is given by:

$$F = nA\frac{dv}{dx} \tag{1}$$

where, 'A' is the area of contact of the layers, 'dv/dx' is the velocity gradient between them and proportionality constant 'n' is known as the co-efficient of viscosity or simply viscosity. The unit of viscosity is 'poise' and 1 poise = 1 g/cm/s.

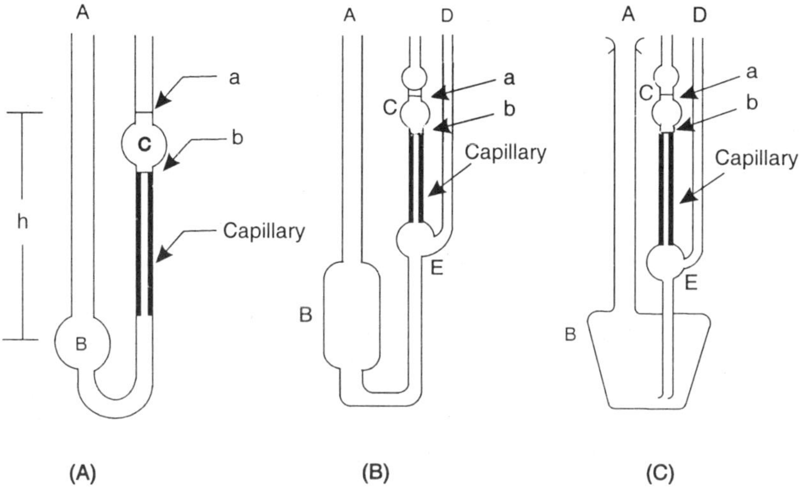

Fig. 3.5: *Different viscometers: (A) Ostwald viscometer, (B) Ubbelhode suspended-level viscometer (USLV), and (C) modified suspended-level viscometer.*

Herman Staudinger estimated the relation between molecular weight and viscosity of a polymer solution as:

$$[\eta] = K\overline{M}_v^{\alpha} \tag{2}$$

where, K and α are constants that depend on the nature of the polymer, solvent and temperatures.

By determining $[\eta]$, i.e. intrinsic viscosity at zero concentration of known $\bar{M}_v$ (molecular weight) using other colligative property, values of K and α are estimated for different polymer/solvent systems. Using these established values of K and α, molecular weights can be determined.

Intrinsic viscosity is defined by,

$$[\eta] = \underset{C\to 0}{\mathrm{Lt}} [\eta_{sp}/C] = \underset{C\to 0}{\mathrm{Lt}} (\eta_{\text{solution}} - \eta_{\text{solvent}})/(\eta_{\text{solvent}} \times C) \tag{3}$$

where, η_{sp} is the specific viscosity of the polymer solution and C is the concentration in g of polymer in 100 ml of the solvent. The specific viscosity η_{sp} is related to the relative viscosity η_r by the equation $\eta_{sp} = \eta_r - 1$, i.e. $\eta_{sp} = (\eta/\eta_o) - 1$. Other terms related to viscosity measurements are given in Table 3.2.

Table 3.2: *Solution viscosity nomenclature*

Name	*Symbol*	*Definition*[a]
Relative viscosity	η_r	$\eta_r = (\eta/\eta_o) = t/t_o$
Specific viscosity	η_{sp}	$\eta_{sp} = \eta_r - 1 = (\eta/\eta_o) - 1$
Reduced viscosity	η_{red}	$\eta_{red} = \eta_{sp}/C$
Inherent viscosity	η_{inh}	$\eta_{inh} = \ln \eta_r/C$
Intrinsic viscosity	$[\eta]$	$[\eta] = \lim_{C\to 0} (\eta_{sp}/C) = (\ln \eta_r/C)_{C\to 0}$

[a]η = solution viscosity; η_o= solvent viscosity; t_o & t = time of flow for solvent and solution

Apparatus: Ubbelhode suspended level viscometer (USLV), constant temperature water bath, thermometer, stop watch, electronic balance and 50 ml standard flasks.

Materials: Polymer, solvent.

Procedure: Prepare a stock solution of polyvinyl alcohol (PVA) by dissolving 2 g of PVA in 50 ml of boiling water. It is then cooled, transferred quantitatively to a 100 ml standard flask and made upto 100 ml to obtain a 2 % PVA solution. From this stock solution, prepare 0.8%, 0.7%, 0.5%, 0.4% and 0.2% PVA solution by pipetting out calculated amount of stock PVA solution and diluting it into 50 ml standard flask. Place 10 ml of the prepared solution into USLV and determine the flow time t_c. Dilute the solution successively with a known amount of the solvent and measure flow time for those concentrations. Record the time of flow for the solvent (t_0) alone and calculate the reduced viscosity $[(t_c - t_0)/t_0]$. Repeat the steps for the constituent polymers. While measuring the viscosity following precaution should be considered. Solvent should be free from dust and colloidal particles. It should be non-toxic. The viscometer should be fixed perpendicular to the working table. It should be immersed in water bath to maintain constant temperature. Same quantity of solution should be take for each analysis.

Draw the plots of reduced viscosity and inherent viscosity (on Y-axis) versus concentration of the polymer solution (on X-axis). A straight line is drawn through the points and extrapolated to the zero concentration. The intrinsic viscosity $[\eta]$ is the intercept of the line at zero concentration. Substitute the values of K and α to equation (2) and calculate the viscosity average molecular weight.

Advantages: It is a simple and easy method. Molecular weight more than one lakhs can be measured. But polymer should be soluble in the commercially available solvents completely.

Observations:

Time of flow for solvent, t_o =.........

Density of solvent = ...

Conc., C (g in 100 ml)	*Flow time, t (sec)*			η_r = η/η_o = t/t_o	η_{sp} = $\eta_r - 1$ = $(\eta/\eta_o) - 1$	η_{sp}/C	*ln* η_r	*ln* η_r/C
	1	*2*	*mean*					

Typical plots of viscosity data for computing intrinsic viscosity [η]. The ordinate intercept gives [η].

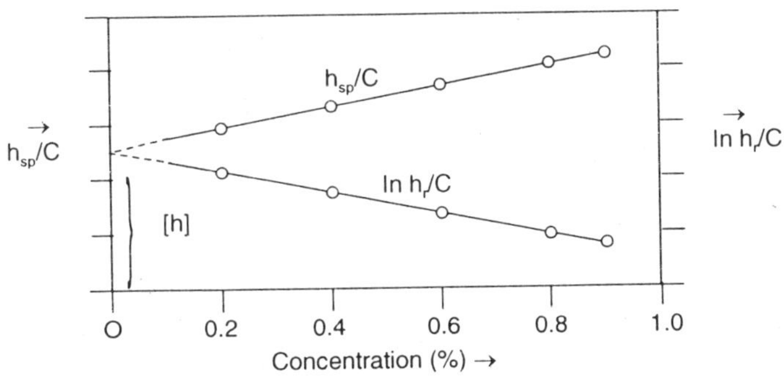

Fig. 3.6: *Plot of η_{sp}/C and ln η_r/C verses concentration of polymer solution*

K= 45.3 × 10^{-3} and α = 0.64 for PVA solution in water. Calculate the $\overline{M}_v$, using equation (2).

(*n*) From the graph = ...

Intrinsic Viscosity and Molecular size

Estimation of root mean square end-to-end distance of polymer chain: Theories of the fractional properties of polymer molecules in solution show that the intrinsic viscosity is proportional to the effective hydrodynamic volume of the molecule in solution divided by its molecular weight. The effective volume is proportional to the cube of a linear dimension of the randomly coiling chain. That means in order to determine the root mean square end-to-end distance $(\bar{r}^2)^{1/2}$, of the polymer chain the following equation is made use of:

$$(\overline{r^2})^{3/2} = [\eta]\, M / \phi \qquad (4)$$

where, ϕ is a universal constant, independent of polymer type, solvents, and temperature. Where $[\eta]$ is expressed in dl/g and the value of ϕ is 2.84×10^{21}. Substitute the values of ϕ, $[\eta]$ and M, in the above equation, to calculate $(\overline{r^2})^{3/2}$.

Hence, root mean square end-to-end distance $(\overline{r^2})^{1/2}$ can be calculated from the cube root of the value of $(\overline{r^2})^{3/2}$. i.e.,

$$(\overline{r^2})^{1/2} = \left\{\left[(\overline{r^2})^{3/2}\right]^{1/3}\right\} \qquad (5)$$

In all other cases the size of the random coil must be expressed interms of statistical parameters such as the $(\overline{r^2})^{1/2}$ its radius of gyration, the root mean square distance of the elements of the chain from its centre of gravity $(\overline{s^2})^{1/2}$. For linear polymers that are not appreciably extended beyond their most probable shape, the mean square end-to-end distance and the square of the radius of gyration are simply related :

$$\overline{r^2} = 6\overline{s^2} \qquad (6)$$

The mean radius of gyration $\overline{s^2}$ is calculated using the equation, $\overline{s^2} = \dfrac{\overline{r^2}}{6}$. (7)

Viscosity average molecular weight constants at room temperature (25 or 30°C), for some of the polymers are given in Table 3.3.

Result: Report the, (i) Viscosity average molecular weight of the given sample.

(ii) Root mean square end-to-end distance $((\bar{r}^2)^{1/2}$

(iii) Square of the radius of gyration $(\bar{s}^2)$.

Experiment 4

Determination of Acid Value and Iodine Value for a Given Polymer Sample

Principle: Acid value and iodine value plays a very important role in selecting suitable carbon black, polymer or filler and

Table 3.3: *Mark-Houwink-Sakurada constants for some polymer-solvent systems (K and α)*

Polymer	*Solvent*	*Temp (°C)*	*Mol. wt ($M \times 10^{-4}$)*	*$K \times 10^2$ (ml/g)*	*α*
Natural rubber	Toluene	25	4–150	5.0	0.67
Polymethyl metha-crylate (PMMA) (atactic)	Chloroform	25	8–140	0.48	0.80
Polymethyl metha-crylate (atactic)	MEK	25	8–140	0.68	0.72
Polystyrene (PS) —atactic	Benzene	20	0.12–14	1.23	0.72
Polystyrene (PS) —atactic	Toluene	20–30	2–200	1.05	0.73
Polystyrene (PS) —atactic	MEK	20–40	0.8–400	3.82	0.58
Polyvinyl alcohol	Water	25	0.85–17	30.0	0.50
Polyvinyl acetate	Methanol	30	2.7–130	3.14	0.60
Polyiso-butylene	Cyclohexane	30	0.05–320	2.88	0.69
Polyiso-butylene	Toluene	25	14–34	8.70	0.56
SBR	Toluene	25	1–10	37.9	0.71

to know the processing characteristic properties. The "*Acid value* which is the number of milligrams of potassium hydroxide required to neutralize one-gram of the polymer sample", which indicates the amount of free acid present.

The "*Iodine value* which is the number of grams of iodine that combines with 100 g of oil or fat", which gives the degree of unsaturation of acids in the substance.

(a) Estimation of Acid Value

Acid value experiment measures the acid and basic constituents present in the compounds that are soluble in mixtures of isopropyl alcohol/water. This experiment reveals that the purity and functionality of the organic compound.

Apparatus: Conical flask, pipette, burette, hot plate cum magnetic stirrer, analytical balance, etc.

Materials: 1:1 isopropyl alcohol/water mixture, 0.1N alchoholic NaOH solution, phenolphthalein indicator, and acid sample.

Procedure: Weigh out a known amount of given polymer sample into a clean conical flask, dissolve it in 50 ml of 1:1 isopropyl alcohol/water mixture. Reflux it till the solution becomes clear. Titrate against 0.1N sodium hydroxide solution using phenolphthalein as an indicator till permanent pale pink colour is obtained. Conduct the experiment in duplicate and report the average value. Acid value can be calculated by using following relation:

$$\text{Acid value} = \frac{V_{\text{NaOH}} \times N_{\text{NaOH}} \times 56.1}{\text{Weight of the sample}}$$

where, V = Volume of NaOH used, N = Normality of NaOH, and 56.1 is the equivalent weight of NaOH.

Result: Report the acid value of the given polymer sample.

(b) Estimation of Iodine Value

Apparatus: 250 ml iodine flask, electronic balance, hot-plate cum magnetic stirrer, burette, microburette.

Materials: Sample, 0.1 N sodium thiosulphate, cabon tetrachloride, iodine monobromide solution, potassium iodide, starch.

Procedure: Weigh about 1 gram of the given sample and transfer it into a clean 250 ml iodine flask. Add 25 ml of carbon tetrachloride and exactly 20 ml of iodine monobromide solution and mix well and keep undisturbed for one hour. Add 10 ml of potassium iodide and titrate against 0.1N sodium thiosulphate using starch as indicator near the end-point (colourless). Repeat the experiment in duplicate. Similarly carryout a blank titration without sample. The iodine value is calculated by using the relation:

$$\text{Iodine value} = \frac{(B - S) \times N_{Na_2S_2O_3} \times 126.9}{(\text{Wt. of the sample}) \times 10}$$

where, B = volume of $Na_2S_2O_3$ for blank titration; S = volume of $Na_2S_2O_3$ for sample titration, N = normality of $Na_2S_2O_3$ and 126.9 is the equivalent weight of iodine.

Iodine value represents the degree of unsaturation of the sample.

Result: Report the iodine value of the given sample.

Experiment 5

Determination of Epoxy Equivalent Weight (EEW) by the hydrogen chloride-acetone method and hydrolyzable chlorine content of an epoxy resins

Introduction: The term epoxide equivalent (or epoxy equivalent weight (EEW)) is defined as the weight of resin in grams, which contains one-gram equivalent of epoxy. The term *epoxy value* is also employed to represent the fractional number of epoxy groups contained in 100 g of resin. If the resin chains are assumed to be linear with no branching and it is further assumed that an epoxy group terminated each end, then the epoxide equivalent is one-half the average molecular weight of a di-epoxy resin, one-third of the average molecular weight of a tri-epoxy system, etc. Dividing the epoxy value by 100 gives the epoxy equivalent. Examples for epoxy resins are diglycidyl ether of bisphenol-A (DGEBA), tetraglycidyl ether of diaminodiphenylmethane (TGDDM), etc. EEW value is also used to calculate the functionality of epoxy resin.

Significance and scope: These are the simple and feasible chemical methods to characterize the epoxy resin. These experiments are used to estimate the purity and functionality of prepared or commercially available epoxy resins. To optimize the synthetic conditions of epoxy resins, EEW and percent hydrolyzable chlorine content values are useful.

(a) Determination of EEW of Epoxy Resins

Apparatus: Electronic balance, 250 ml iodine flask, water bath, microburette or pipette, etc.

Materials: Epoxy resin, hydrogen chloride-acetone mixture (2.8 ml of 36.5% HCl in 140 ml of acetone); 0.1 N sodium hydroxide solution, methyl red indicator (0.1 g methyl red per 100 ml methyl alcohol), methanol.

Procedure: Place about 0.3 g of epoxy resin into a 250 ml iodine flask. Pipette out 25 ml of hydrogen chloride-acetone solution mixture. Swirl the solution until the entire sample has been dissolved. After the sample has dissolved, wash the stopper with a little acetone and collect all the washings into same flask. Add few drops of methyl red indicator. Titrate with 0.1N sodium hydroxide. Similarly, carry out blank titration without epoxy resin sample. The reaction of epoxy with HCl is given below:

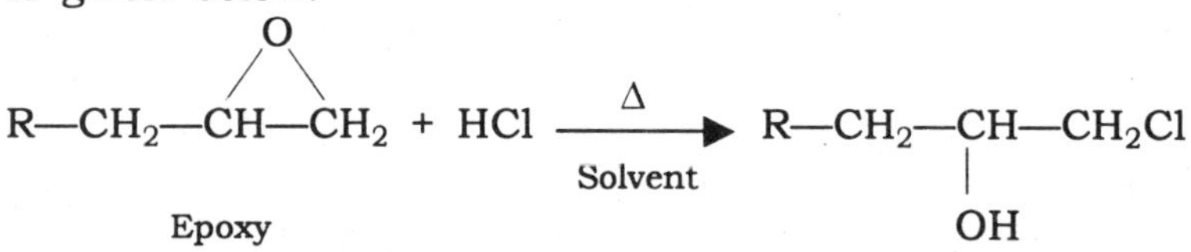

Epoxy equivalent weight can be calculated using the relation given below:

$$\text{Epoxy equivalent weight} = \frac{1000 \times (W_2 - W_1)}{N_{\text{NaOH}} \times (V_1 - V_2)} \text{ g/mol}$$

where, $W_2 - W_1$ = Weight of epoxy resin taken; N_{NaOH} = normality of sodium hydroxide; V_1 = volume of NaOH for blank titration, V_2 = volume of NaOH for sample titration.

(**Note:** Standardize the NaOH solution by using standard solution of potassium hydrogen phthalate and phenolphthalein as an indicator).

Result: Report the epoxy equivalent weight of the given epoxy resin.

(b) Detezrmination of Hydrolyzable Chlorine Content of Epoxy Resins

Epoxy resins synthesized via epichlorohydrine route will contain atleast some hydrolyzable chlorine (i.e., present as 1,2-chlorohydrine as a result of incomplete dehydrohalogenation). This chlorine is considered as *active*. The active chlorine is capable, of blocking the reactions with Lewis base catalysts, and when the amount of the catalyst is fairly critical, variations in hydrolyzable chlorine can result in significant variations in cured properties of epoxy resin.

Apparatus: Electronic balance, water bath, water-cooled condenser, 250 ml iodine flask, burette, 25 ml pipette, 10 ml measuring jar.

Materials: Epoxy resin, 0.5 N alcoholic KOH (dissolve 28.05 g of reagent-grade KOH in 1 liter of methanol), 0.5 N hydrochloric acid, toluene, ethanol, phenolphthalein indicator.

Procedure: Introduce 1 to 2 g of epoxy to be tested into each of two 250 ml iodine flasks. Pipette out 50 ml of the 0.5 N alcoholic KOH into each sample and into two blanks. Add 15 ml of toluene from a graduated cylinder into each flask. Stopper it and swirl until resin is dissolved and add 3 or 4 glass beads. Connect the flask to reflux condenser, and place it on hot plate. Reflux for 15 minutes and cool the mixture to room temperature with condenser in place. After cooling, wash down the condenser with 5 ml ethyl alcohol and collect all washings in the same flask. Add 2–3 drops of phenolphthalein indicator. Titrate with 0.5N hydrochloric acid to the first disappearance of the pink colour. Percent hydrolyzable chlorine content can be calculated using the following relation :

$$\text{Hydrolyzable chlorine content} = \frac{(B - S) \times N_{HCl} \times 3.55}{\text{Weight of the epoxy resin taken}} \%$$

where, S = millilitres of acid required to titrate sample; B = millilitres of acid required to titrate blank; N_{HCl} = normality of HCl and 35.5 is the equivalent weight of chlorine (3.55 is after conversion).

Result: Report the percent hydrolyzable chlorine content of the epoxy resin.

Experiment 6

Determination of surface area of carbon black by iodine adsorption method.

Introduction: Characteristic properties of carbon black are: (i) particle size/surface area, (ii) structure, (iii) aggregate complex, (iv) surface activity, and (v) chemical nature of the filler.

Principle: The iodine adsorption number and cetyl trimethyl ammonium bromide (CTAB) are useful methods in estimating the surface area of carbon black in liquid phase. From the surface area value we can distinguish different grades of carbon black (Tables 3.4 and 3.5).

(*i*) **Iodine number (ASTM D 1510):** Weight in milligrams of iodine adsorbed per gram of carbon black. Higher the iodine number higher is the surface area. Basically it is the measure of external surface area.

(*ii*) **Cetyl trimethyl ammonium bromide (CTAB):** Measures the external area only (ASTM D 3765–89). CTAB is excluded from the smaller interstices and thus represents the portion of the particle surface area in contact with or available for the polymer.

ASTM designation of rubber grade carbon blacks are as follows:

(*i*) The system consists of a letter, followed by three digits. The letter classifies the cure rate, the first

Table 3.4: *Classification of different grades of carbon black*

Description of black	*Old classi-fication*	*ASTM classi-fication*	*Typical particle dia (in nm)*
0	–	–	1 - 10
Super abrasion furnace black	SAF	N110	11 - 19
Intermediate super abrasion furnace black	ISAF	N220	20 - 25
High abrasion	HAF	N330	26 - 30
Fast extrusion furnace	FEF	N550	39 - 58
General purpose	GPF	N660	40 - 48
Semi-reinforcing furnace	SRF	N770	49 - 60
Fine thermal black	FT	N880	101 - 200
Medium thermal black	MT	N990	201 - 350
Easy processing channel black	EPC	S300	21

***Table 3.5:** Some of the typical properties of carbon black*

Industry Name	*ASTM No.*	*Iodine No. (mg/g)*	*Nitrogen (mg/g)*	*DBP Absorption (cc/100g)*	*CTAB Absorption (m^2/kg)*	*Pour Density (kg/m^3)*
SAF	N110	145	143	113	126	335
ISAF LS	N210	118	120	78	113	–
ISAF	N220	121	119	114	111	345
ISAF LM	N231	121	117	92	108	390
ISAF HS	N234	120	126	125	119	320
HAF LS	N326	82	84	72	83	465
HAF	N330	82	83	102	83	375
HAF LM	N339	90	96	120	95	345
HAF HS	N347	90	90	124	88	335
FEF	N550	43	42	121	42	360
GPF HS	N650	36	38	122	38	370
GPF	N660	36	35	90	35	425
SRF LS	N762	27	28	65	27	505
SRF HS	N765	31	31	115	33	375
SRF HM	N774	29	29	72	29	495

digit indicates the typical particle size and the last two digits second digit structure are assigned arbitrarily to distinguish blacks within a given particle group.

(*ii*) The letters used to designate cure rate are N and S. The letter N stands for *normal cure rate* and S stands for *slower cure rate*.

(*iii*) The first number recognizes the particle size as a fundamental characteristic of carbon blacks. The particle size range of rubber grade carbon blacks has been divided into 10 groups based on the partice size and each group has been assigned a digit to describe the group.

(*iv*) The particle size is expressed in nanometres. The groups are not equal in numerical size, but are chosen so that blacks now known as SAF, ISAF and HAF have not fallen in the same classification.

(*v*) The last two numbers would indicate the differences within a group. This difference can be in structure level, modulus level or any of the physical/chemical properties associated with a particular black. Thus, N330 and N347 will indicate two blacks having a normal type cure rate with particle size of 26–30 nm with different sturcture.

Apparatus: 250 ml iodine flask, centrifuge, electronic balance.

Materials: Carbon black, 0.1N iodine solution, 0.1 N sodium thiosulphate, starch indicator.

Procedure: Weigh around 1 g of carbon black and transfer it in to a clean 250 ml iodine flask. Add 25 ml of 0.1N iodine solution and shake well for around 3–5 minutes. Centrifuge the solution and pipette out 10 ml of carbon black free iodine solution-into a conical flask and titrate it against 0.1 N sodium thiosulphate solution using starch as indicator near the end point. Similarly, carryout blank titration without sample.

Calculation: Surface area of carbon black can be calculated as:

$$\text{Surface area of carbon black} = \frac{(B - S) \times V_B \times N_{Na_2S_2O_3} \times 126.9}{B \times (W_2 - W_1)}$$

where, B is the volume of $Na_2S_2O_3$ of blank titration; S is volume of $Na_2S_2O_3$ of sample titration; V_B is volume of the iodine taken for titration; $(W_2 - W_1)$ is weight of the carbon black; $N_{Na_2S_2O_3}$ is the normality of sodium thiosulphate and 126.9 is the equivalent weight of iodine.

Result: Report the surface area of carbon black. Compare the experimental value with the literature value and name the carbon black.

Experiment 7

Determination of D.B.P. value and Sieve analysis of Carbon black

(i) DBP Adsorption Value (ASTM D-2414) of Carbon Black

High structure = High number of particles/aggregates = High Modulus and better processability.

Structure is measured by the following method:

Method: The volume of dibutyl phthalate (DBP) in cubic centimetres absorbed by 100 g of carbon black. The fluid fills up void spaces in carbon black aggregate, when all is filled up by the DBP, carbon black mass experiences a sharp rise in the viscosity. DBP value is defined as the "number of moles of DBP required to mix 100 g of carbon black thoroughly".

Apparatus: Petridish, analytical balance (accuracy 0.001 g), and microburette.

Materials: Carbon black, dibutyl n-phthalate (DBP).

Procedure: Take a known amount ($\simeq$ 5g) of carbon black in a petridish. Add dibutyl n-phthalate drop wise from a microburette. Thoroughly mix the carbon black with DBP. Add DBP until carbon black is completely wet. Note down the volume of DBP run down from the burette. Calculate the DBP value of carbon black using the following formula:

$$\text{DBP value} = \frac{100 \times \text{Volume of DBP rundown}}{\text{Weight of carbon black taken}}$$

Result: Report the DBP value of the given carbon black

(ii) Grit Analysis (ASTM D-1921) of Fillers

Theory: The particle size and particle size distribution are important since these two characteristics of materials have a great effect on compounding, processing and bulk handling and property of the final product. Large and fairly uniform particles are easier to handle and process, whereas fine particles are difficult to process. Sieve analysis is used to determine the shape and size of the carbon black particles. In this analysis filler is allowed to pass through meshes of various specification.

High surface area = High reinforcement = Small particle size

"Particle size can be stated as, it is inversely related to surface area and it is a microproperty. Electron microscope is used to measure the particle size of carbon blacks."

Apparatus: Meshes of different but known numbers, analytical balance.

Materials: Fillers

Procedure: The test method for the determination of particle sizes and their distribution (ASTM D-1921) employs a sieve analyzer consisting of sieves of decreasing opening size placed vertically in a series along with a shaker to facilitate separation of various sized particles. About 10 g of filler is poured from the top and allowed to pass through the sieves. Then shake the sieves simultaneously and vigoursly in such a way that the carbon black particles pass through successive meshes. If required run water forcefully so that the particles are forced through the mesh. Then dry the collections in each sieve separately. The particle distribution is ascertained by weighing the collections in each sieve. Calculate the sieve residue using the formula given below:

$$\text{Pecentage of sieve residue} = \frac{\text{Weight of residue}}{\text{Weight of filler black taken}} \times 100$$

Result: Report the percentage sieve residue in each mesh with the particle size of the given filler.

Experiment 8

Determination of Concentration of Formaldehyde in formaline solution

Method: Formaldehyde may be estimated in solution by oxidizing it into formic acid by means of a known quantity of iodine dissolved in an excess of NaOH solution. The oxidation is, in effect, carried out by the sodium hypoiodite formed by the action of the iodine on the alkali, the formic acid being then neutralized by the alkali present.

$$I_2 + 2NaOH \rightarrow NaI + \underset{\text{Sodium hypoiodite}}{NaIO} + H_2O$$

$$\underset{\text{Formaldehyde}}{HCHO} + NaIO \rightarrow \underset{\text{Formic acid}}{HCOOH} + NaI$$

When the oxidation is complete, the excess of hypoiodite is estimated by acidifying the solution, and then titrating the iodine, thus, formed against standard sodium thiosulphate solution. It will be seen that each molecule of iodine used in the original oxidation corresponds to one molecule of formaldehyde (One molecule of formaldehyde consumes one molecule of iodine).

$$NaIO + NaI + 2HCl \rightarrow 2\ NaCl + H_2O + I_2 \uparrow$$

Apparatus: 25 ml pipette, burette and conical flask.

Materials: Formaline solution, M/10 iodine solution, 10% NaOH solution, M/10 $Na_2S_2O_3$ solution and starch as an indicator.

Procedure: Pipette out 25 ml of the given formaline solution and 50 ml of M/10 iodine solution into a clean conical flask. Now, add 10% sodium hydroxide solution dropwise until the liquid becomes pale-yellow in colour, and allow the solution to stand, with occasional shaking, at room temperature for at least 10 minutes. Then acidify with dilute hydrochloric acid (free from chlorine) in order to liberate the remaining iodine. Titrate the latter with M/10 sodium thiosulphate solution, using starch as an indicator near the end-point. Weight per litre of formaldehyde can be calculated using the following relation:

1 ml of M/10 of iodine solution = 15 mg of formaldehyde

Result: Report the weight/litre of formaldehyde.

Experiment 9

Determination of Aniline Point and Specific Gravity of Processing Oil

Petroleum oil and plasticizers are complex mixtures of individual molecules containing various ratios of unsaturated aromatic rings, saturated rings (known as naphthenic rings) and paraffinic side chains. The oils are then classified as aromatic, naphthenic and paraffinic according to the ratios of the three types of structure present, schematically illustrated in Table 3.6.

It should, however, be made clear that an oil designated as aromatic will not be totally aromatic, with similar comments for the so-called paraffinic and naphthenic oil. It has been found possible to carryout a carbon-type analysis using a correlatio relating to the carbon-type distribution to the viscosity-gravity constant and the refractivity intercept. Viscosity-gravity

Table 3.6: *Typical properties of petroleum oils used in rubber industries*

Properties	*Paraffinic*	*Naphthenic*	*Aromatic*
Viscosity (cSt)			
At 40°C	19.7	110.2	763.5
At 100°C	4.0	8.0	17.0
Specific gravity @ 15°C	0.861	0.932	1.018
VGC	0.8090	0.8850	0.9800
Refractive index	1.4751	1.5167	1.5804
Aniline point (°C)	96.0	75.0	38.2
Molecular-type analysis (%)			
Aromatics	12.1	42.8	80.0
Saturates	87.5	54.4	12.2

constant (VGC) is a measure of aromaticity calculated from the equation;

$$(VGC) = 10\,G - 1.0752\,\log(V - 38)/10 - \log\,(V - 38)$$

where, G is the specific gravity at 15°C and V is the Saybolt viscosity at 38°C.

When selecting oil there are five main factors to take into account:

(*i*) The viscosity gravity constant—a measure of aromaticity,
(*ii*) the oil viscosity—a function of both structure and molecular weight,
(*iii*) presence of heterocyclics and other materials containing nitrogen, oxygen or sulphur,
(*iv*) oil loading and
(*v*) cost.

(a) Determination of Aniline point

"**Aniline Point** is the temperature at which the two immiscible liquids or oil and aniline mix without any line of demarcation". In other words, it is the temperature at which oil just becomes completely miscible with equal volume of aniline.

This is a convenient method at accessing aromatic nature of oils because aromatic hydrocarbons have much lower aniline point than paraffins and naphthalenes. Consequently, a low aniline point indicates high aromatic contents (aromaticity) and low cetane number. Aniline point is determined by mixing of equal volumes of oil and dry distilled aniline in a test tube provided with a stirrer and thermometer. The mixture is slowly warmed with stirring until complete miscibility is attained. Then the mixture is allowed to cool slowly until separation just occurs, lowest temperature at which complete miscibility disappears is termed as *aniline point.*

Apparatus: Thermometer, test tube, measuring jar, Bunsen burner, stand, etc.

Materials: Aniline, oil (sample).

Procedure: The processing oil is taken in a clean dry test tube and same amount of aniline is added to the same test tube. There will be a clear line of demarcation visible between the two liquids. The thermometer is fitted in to the test tube with the help of a cork. The test tube is mounted properly to the stand. Now, the test tube is heated slowly with Bunsen burner. The rise in temperature should be carefully observed. As the two liquids mix properly, the temperature is noted. The experiment is repeated by cooling and re-heating for two to three times for same sample to get the concordant values and report the average value.

Result: Aniline point of the oil = ... °C.

(b) Determination of Specific Gravity

Specific gravity is defined as the "ratio of the weight of sample to the weight of water of same volume. It has no units because it is the ratio between two weights. It is one of the simple physical properties of the matters.

Apparatus: Electronic balance, specific gravity bottle (10 ml).

Materials: Oil and acetone.

Procedure: A clean, dry specific gravity bottle is weighed (W_1 g). The specific gravity bottle is filled with distilled water and weighed again (W_2 g). Now the specific gravity bottle is emptied and dried using acetone. Now, it is filled with the oil supplied. The weight of the specific gravity bottle with processing oil is recorded as (W_3 g).

Specific gravity of oil can be calculated as:

$$\text{Specific gravity} = (W_3 - W_1) / (W_2 - W_1)$$

Result: Report the specific gravity of oil.

Experiment 10

Determination of Concentration of Hydrogen Peroxide

Hydrogen peroxide is usually encountered in the form of an aqueous solution containing about 6, 12 or 30 percent hydrogen peroxide, and frequently referred to as 20, 40 and 100 volume hydrogen peroxide respectively. This terminology is based on the volume of oxygen liberated, while boiling the solution. Thus, 1 ml of 100 volume hydrogen peroxide will yield 100 ml of oxygen measured at STP.

The following reaction occurs when potassium permanganate solution is added to hydrogen peroxide solution acidified with dilute sulphuric acid:

$$2MnO_4^- + 5H_2O_2 + 6H^+ \rightarrow 2Mn^{2+} + 5O_2 + 8H_2O$$

Significance: This method is useful to check the purity and concentration of H_2O_2.

Apparatus: Pipette, conical flask and burette.

Materials: H_2O_2, dilute sulphuric acid and potassium permanganate.

Procedure: Pipette out 25 ml of given hydrogen peroxide into a clean 500 ml conical flask, dilute with 200 ml water, add 20 ml of dilute sulphuric acid (1:5), and titrate with standardized potassium permanganate to the first permanent pale-pink colour. Repeat the titration till the concordant titre values are obtained. Normality of H_2O_2 can be calculated as:

$$\text{Normality of } H_2O_2 = \frac{N \times V \text{ of } KMnO_4}{V \text{ of } H_2O_2}$$

where, (N × V) of $KMnO_4$ are the normality and volume of $KMnO_4$; V of H_2O_2 is the volume of hydrogen peroxide.

$$\text{Wt/litre of } H_2O_2 = \text{Normality of } H_2O_2 \times \text{Eq. Wt. of } H_2O_2$$

The equivalent weight of hydrogen peroxide is 17.

Result: Report the concentration of H_2O_2.

Experiment 11

Determination of the Saponification Value of a given Oil

Principle: The hydrolysis of an ester with the formation of an alcohol and an acid when the process is carried out in an alkali solution is called *saponification*. It can be defined as "number of milligrams of KOH required to completely hydrolyse (saponify) one gram of oil or fat."

In practice a known amount of an oil or fat is refluxed with excess amount of standard alcoholic potash solution and the unused (excess) alkali is titrated against the standard acid. The rection is as follows:

$$
\begin{array}{lcccccc}
CH_2OCOR & & & & CH_2OH & & \\
| & & & & | & & \\
CHOCOR & + & 3KOH & \longrightarrow & CHOH & + & 3RCOOK \\
| & & & & | & & \\
CH_2OCOR & & & & CH_2OH & & \\
\text{Oil or fat} & & & & \text{Glycerol} & & \text{Soap}
\end{array}
$$

This technique can also be used to get alkali free soaps.

Apparatus: Conical flask, reflux condenser, burette, pipette, analytical balance, etc.

Materials: Oil, 0.5 N alcoholic KOH, 0.5N HCl and phenolphthalein.

Procedure: Take about one gram of oil into 250 ml conical flask. To this, pipette out 50 ml 0.5N of alcoholic KOH. Reflux this mixture for 20–30 minutes. Cool and titrate against 0.5 N HCl using phenolphthalein as an indicator. Similarly, carryout blank titration without the sample. Repeat the procedure until concordant values are obtained. Saponification value of the sample can be calculated by using the following equation :

$$\text{Saponification value} = \frac{56.1\,(B - S) \times N_{HCl}}{\text{Weight of sample}}$$

where, B is the volume of the blank titration; S is the volume of the sample titration and N_{HCl} is the normality of HCl.

Result: Report the saponification value of the given oil.

Experiment 12

Determination of the Flash, Fire and Pour Point of a given Oil

Criteria for selection of oils: In India, the Institute of Petroleum authoritatively assesses the nature and quality of oils. Specific gravity, viscosity, calorific value, flash point, pour point, cloud point, sulphur content, distillation range, carbon residue, ash content, etc., are some of the characteristic properties of oils.

Oils used for burning purpose must have high calorific value, low viscosity and low pour point. While oils used for heat treatment (oil quenching) must have high distilling range, low carbon and ash content and less viscous, while oils used for lubrication purposes must have low pour point, ash content and carbon residue.

(a) Flash point: The flash point of an oil is the lowest temperature to which an oil is to be heated until sufficient inflammable vapour is driven from it which produces a flash when it brought into contact with a flame. This is a measure of fire risk while handling in bulk quantity. Permissible conditions of storage and use are different for products with flash points below 35°C (gasoline) and between 35°C and 70°C (kerosene) and above 70°C (diesel oils).

Oils with flash points less than 35°C including gasoline are regarded as dangerous and highly inflammable for transportation and storage purposes. Fuels above 70°C flash values are regarded as safe. Two different types of instruments are available commercially; Abel's tester for flash points less than 60°C (kerosene ≃ 50°C) and Pensky-Marten tester for those fuels with values above 60°C (gas oil, diesel, fuel, oil etc., which have a flash point between 75°C and 120°C).

(b) Fire point: Lowest temperature of an oil at which sufficient vapours are formed above oil surface and vapours catch fire and burn continuously when brought in contact with flame.

Since sufficient vapours are required for continuous burning, fire point is always higher than flash point. Flash and fire points of oils/fuels can be determined by Abel's and Pensky-Martin's apparatus.

Determination of flash and fire points of oil

(1) By Abel's-Pensky method for light oil Apparatus:

The apparatus consists of an oil cup provided with a cover containing parts for introducing the burning splinter (Fig. 3.7). The cup is surrounded by water bath which is heated. The oil gets heated through the water bath. A stirrer is provided for stirring oil to get uniformity in temperature. There is a small hole to introduce thermometer in the oil cup for measuring the temperature.

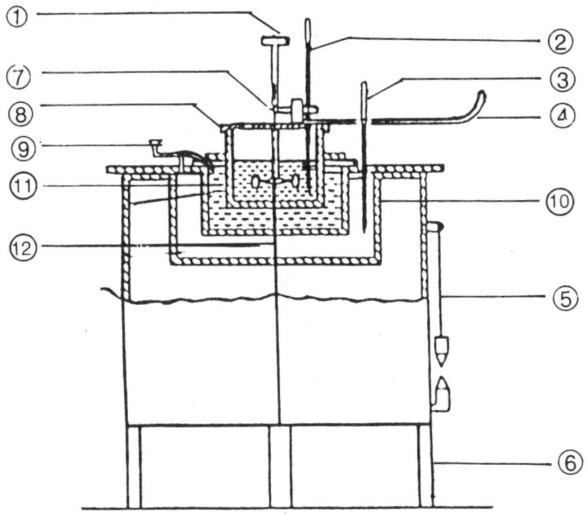

1. Stirrer
2. Thermometer
3. Thermometer
4. Lever to open apperture
5. Plumb line
6. Iron stand
7. Opening
8. Cover plate
9. Water inlet
10. Doubled jacketed brass vessel
11. Brass oil cup
12. Air cavity

Fig. 3.7: *Abel's Flash Point Apparatus*

Material: Light oil.

Procedure: Whole setup is mounted in a water bath with annular space between the two. Oil is filled up to the marked level and temperature is slowly raised by means of water bath. Rate of temperature rise depends on the type of oil, (for oils of flash less than 45° C heating rate must not exceed 2°C/ min). At regular intervals of time, the test flame is brought near the surface of oil and temperature of oil when a flame produces a flash is recorded as flash point.

To determine the fire point, continue heating so that the sample temperature increases at the rate of 5°/min. Continue the application of test flame at 2°C intervals until the oil ignites and continues to burn for atleast 5 seconds. Record the temperature at this point as the fire point of the oil.

2. By Pensky-Marten's Flash point Apparatus

Pensky-Marten's apparatus is used for fuels. The flash values of which are above 60°C, i.e. for majority of fuels. Apparatus is similar to Abel's and heated by air bath (that means the apparatus is heated electrically using the hot plate).

Scope: Flash and fire point is measured as per IP-34 and IS 1448 (P-21) 1970 respectively. This method covers the determination of the flash point by Pensky-Marten's closed cup tester of fuel oils, lubricating oils, viscous materials and suspension of solids. This method is not applicable to drying oils, solvent-type liquid waxes or cut-back asphalt's.

Outline of method: The sample is heated in a test cup at a slow and constant rate with continuous stirring. A small test flame is directed into the cup at regular intervals of temperature with simultaneous interruption of stirring. The flash point is taken as the lowest temperature at which the application of the test flame causes the vapour above the sample to ignite momentarily.

Procedure: 1. Thoroughly clean and dry all parts of the cup and its accessories before starting the test, being sure to remove any solvent which had been used to clean the apparatus. Support the tester on a level, steady table. Fill the cup with the sample to be tested to the level indicated by the filling mark. Place the lid on the cup, and set the latter on the electrically heated plate. Insert the thermometer. Light the test flame and adjust it to 4.0 mm in diameter. Apply heat at such a rate that the temperature recorded by the thermometer increases neither less than 5°C nor more than 6°C per minute. Turn the stirrer 90–120 rev/min, stirring in a downward direction.

2. If the sample is known to have a flash point of 105°C or below, apply the test flame when the temperature of the sample is a whole number not higher than 17°C below the flash point, and thereafter record at each degree rise of temperature.

Apply the test flame by operating the mechanism on the cover which controls the shutter and test flame burner so that the flame is lowered into the vapour space of the cup in 0.5 second, left in its lowered position for one second, and quickly raised to its high position. Do not stir the sample while applying the test flame.

3. If the sample is known to have a flash point above 105°C, apply the test flame in the manner prescribed in step 2 at each temperature, that is, a multiple of 30°C, beginning at a whole number temperature reading not higher than 17°C below the flash point.

4. Record the flash point of the temperature with the help of thermometer at different intervals of time and test flame application causes a distinct flash in the interior of the cup. Do not confuse the true flash point with the bluish halo which sometimes surrounds the test flame at applications preceeding the one that causes the actual flash. Continue the test flame to get fire point.

1. Flexible shaft for stirrer
2. Lever arrangement
3. Brass cup
4. Thermometer to measure oil temperature
5. Slit arrangement
6. Hot plate
7. Electrical connection

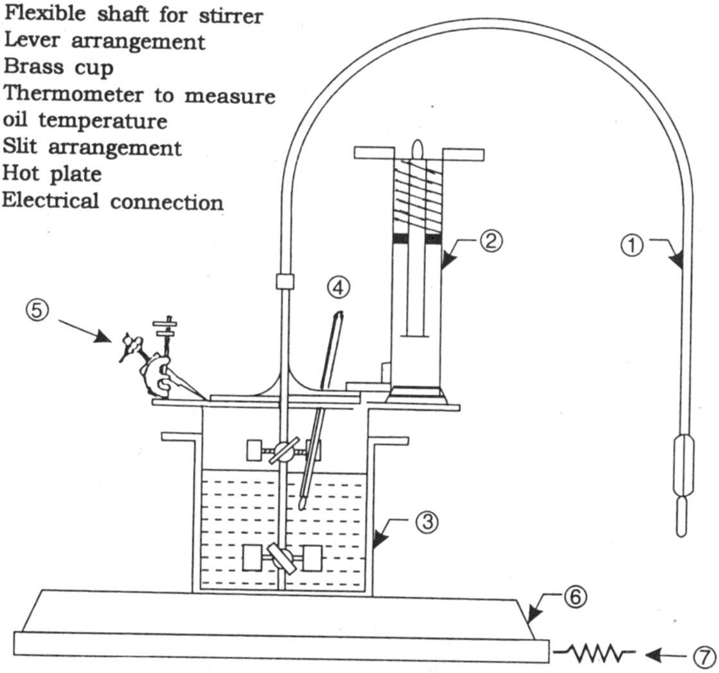

Fig. 3.8: *Pensky-Marten's Apparatus.*

(3.) By Cleveland Open Cup Method

Scope: This method describes the procedure for determining the flash and fire points of petroleum products, except fuels, oils and those products having an open-cup flash point below 70°C.

Apparatus: Cleveland open-cup tester, thermometer, gas cylinder.

Materials: Oil

Procedure: Using Cleveland open-cup tester.

1. Place the flash and fire point setup on a horizontal surface.
2. Wipe out the oil cup with cotton waste to remove any oil residues left in the cup from previous test. If any carbon residues are deposited they must be removed using steel wool. Oil cup must be dried.
3. Support the thermometer in a vertical position with the cup and locate a point half way between centre and side of the cup.
4. Fill the oil cup with the given oil so that the top meniscus is exactly up to the filling line. If any excess is added, it must be removed using dropper. Remove any air bubble on top surface of oil. If any sample is sticking on the outer side of the apparatus, wipe it off.
5. Light the test flame and adjust its intensity to that of the pilot flame.
6. Apply heat (electrically heated using the hot plate) to the oil so that the temperature rises at about 2–3 °C/min. and stirred at 5 rps.
7. When the temperature is about 30 °C below the expected value apply test flame across the oil surface. Test flame is brought to the vapour space at each 2°C reading until flash occurs. The centre of the flame shall move in a plane not more than 2 mm above the plane of the upper edge of the cup passing in one direction first, then in the opposite direction in the next time. The time consumed in passing the test flame across the cup shall be about 1 second.
8. Record the flash point as temperature on the thermometer when a momentary flash occurs on the surface of the oil.
9. Continue heating of the oil in the cup.

10. Record fire point as the temperature read on the thermometer when vapours above the surface ignites and continues to burn continuously for about 5 seconds.
11. Switch off the heat source and allow the oil to cool below its flash point.
12. Duplicate the result by carrying out one more trial.

Pour point: It is the lowest temperature at which oil will not flow when chilled without disturbance.

Apparatus: Apparatus consists of a test tube one and quarter inch in diameter, carrying a thermometer and mounted in a wider tube (to provide air space), which in turn is immersed in a freezing mixture.

Procedure: Oil sample is placed in a test tube and warmed to 60°C to ensure that any waxy crystals present dissolves. Oil is slowly cooled at an interval of 2°C/minute. The tube is removed from the bath for every 5°C fall and held slightly tilted. Ultimately, a point is reached when the tube can be held horizontally for 5 seconds without any movement of oil. The pour point is taken as the temperature 2°C above this solidification temperature.

Result: Report the flash, fire and pour point of the sample.

Experiment 13

Determination of the Percentage of Carbon Black present in the Polymer

Significance: Carbon black is extensively used in polymer industry as reinforcing filler. The method adopted to determine the percentage of carbon black is ASTM D 1603. In polyolefin product like pipe, films, etc. when intended for outdoor applications, carbon black can be useful as a photostabilizer. Gravimetric measurement after pyrolysing the polymer, is made for its quantitative determination in ethylene polymers. The method is useful for quality control. It is suitable when only carbon black is present in the polymer as a non-volatile component at the test temperature.

Apparatus: Combustion boat, furnace, tongs, desiccator and analytical balance.

Materials: Polymer sample, N_2 gas cylinder.

Procedure: The given polymer sample is taken in a well-cleaned, and dried combustion boat. The weight of the boat is measured and designated as W_1 g. Calculated amount of sample with the boat will be W_2 g. The boat containing the sample is placed inside furnace with the help of tongs. The temperature of furnace is maintained at 600°C. The sample is combusted under an inert atmosphere, which is achieved by passing dry nitrogen gas at the rate of 1.7 ml/min. or A commercially available carbon black content apparatus is shown in Fig. 3.9.

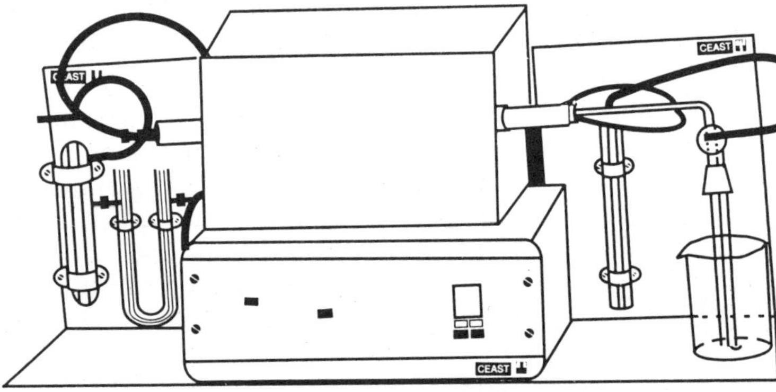

Fig. 3.9: *Apparatus for determining carbon black content in polyolefins.*

The combustion period is about 15 minutes. During combustion, the polymer goes off as pyrolysed product, leaving the carbon black residue. The boat containing the residue is kept inside the desiccator for cooling. Once again the weight of boat is measured and denoted as W_3 g. Calculate the percentage of carbon black by using the relation given below:

Percentage of carbon black = $(W_3 - W_1)/(W_2 - W_1) \times 100$

Result: Report the carbon black content in percent.

Experiment 14

Determination of Viscosity of Oil using Saybolt's and Redwood Viscometers

(a) Estimation of Viscosity by Saybolt's Viscometer

Introduction: Oils have been classified as: gas oil (D), light oil (A), medium oil (E), heavy oil (F) and extra heavy oil (G) according to their viscosities. The apparatus in general use for determination of viscosity is Saybolt's viscometer and viscosity of the oil is expressed in Saybolt seconds and at temperature t °C at which test is made. This type of instrument measures the kinematic viscosity of fluids.

Principle: The Saybolt's viscometer measures the viscosity in empirical units and not in absolute units such as centistokes. It is possible to convert the viscometer readings into absolute units. The method is primarily applicable for viscosity determination of oil, which follows a Newtonian manner. It possesses a linear relationship between shear stress and rate of shear.

Apparatus: Saybolt's viscometer, stop clock, heater and thermometer.

Saybolt's Viscometer: The apparatus consists of an oil cylinder of 3 cm in diameter and 9 cm in depth made of copper. This cylinder is surrounded by water jacket in which water is placed and heated electrically. The oil cylinder has an orifice at the bottom and this orifice is opened and closed by means of a cork valve. The thermometer is placed in the oil bath by which the temperature of the oil bath is recorded (Fig. 3.10).

Materials: Oil.

Procedure: Clean the Saybolt's viscometer. Fill the viscometer with cold water such that water level is more than oil cup indicator level. Place the viscometer on the stand, keep the heater pin and thermometer in the place provided in the instrument. The cork is tightened and the given amount of oil and some excess (65 ml) is filled inside the barrel and the lid is closed. Now, the cork is opened and the stop clock is started. The time taken for 60 ml of oil to flow out is noted. The temperature of the oil is raised and the time taken to flow of oil is determined at constant regular intervals of 10°C. Viscosity at various temperatures are tabulated in centistokes. The graph of Saybolt seconds verses temperature is plotted.

Conversion from Saybolt seconds to centistokes is given below:

$$\text{Centistokes} = 0.22 \times t - 180/t \text{ for } T < 50$$

$$\text{Centistokes} = 0.22 \times t - 195/t \text{ for } 50 < T < 100$$

$$\text{Centistokes} = 0.22 \times t - 135/t \text{ for } T > 100$$

where, t = Saybolt seconds and T = Temperature in °C

Result: Report the viscosity of the sample in centistokes.

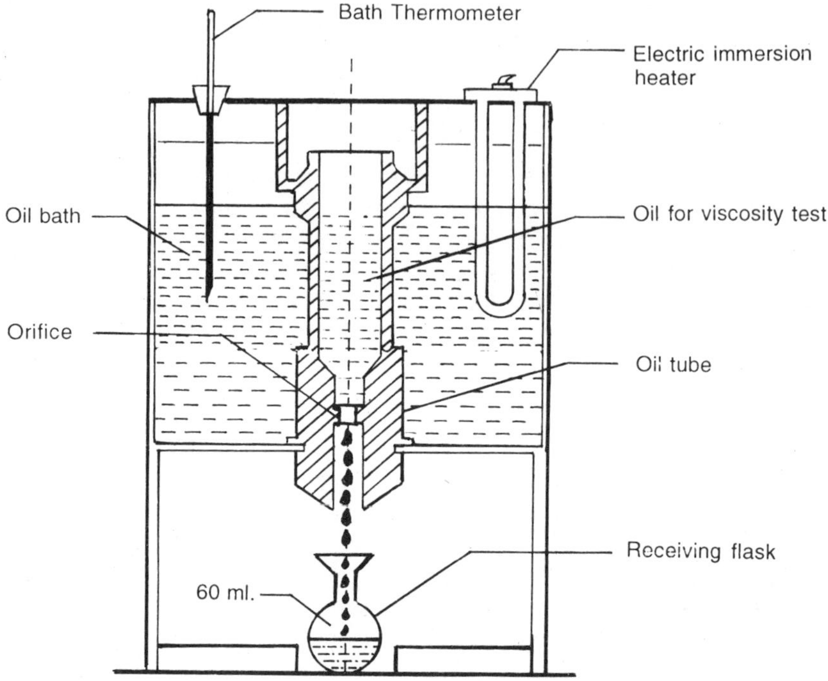

Fig. 3.10: *Saybolt's Viscometer*

(b) Determination of viscosity by Redwood Viscometer

This apparatus is used for determining the viscosity of oil, which is measured by passing the oil through specified hole made in a metallic piece. The viscosity is expressed as a time of flow in seconds.

Scope: Redwood apparatus measures the viscosity in empirical units and not in absolute units such as Centistokes. It is possible to convert Redwood viscometer readings to absolute units, for which the specifications IP70/68 issued by the Institute of Petroleum, London, may be consulted. The method is primarily applicable for viscosity determination of oil, which flows in a Newtonian manner that is, if it possesses a linear relationship between shearing stress and rate of shear under the test conditions.

Apparatus: Clean the oil cup with a suitable solvent, e.g. carbon tetrachloride and dry it thoroughly using soft tissue paper or some similar material which will not leave any fluff. Clean the jet hole with any fine thread.

Setup the viscometer, using the circular spirit level to ensure that is in level. Fill the bath with water for determinations at 93°C and below, and for higher temperatures, with oil having a suitably low viscosity at the test temperature.

Procedure: Heat the viscometer bath to a few degrees above the desired test temperatrure. Pour the prepared sample into the oil cup through a filter of metal gauge not coarser than BS 100 mesh (152/U). Adjust the temperature of the bath until the sample in the cup is maintained at the test temperature by stirring. Stir the sample during the preliminary period, e.g. by means of the ball valve, closing the bottom of the jet by suitable means, but do not stir the sample during the actual determination. When the temperature of the sample has become quite steady at the desired value, adjust the liquid level by allowing the sample to flow out until the surface of the sample touches the filling point. Place the oil cup, and swing the oil cup thermometer towards the closed end of the curved slot. Place the clean and dry stand and a 50 ml flask centrally below the jet with the top of the neck, a few millimetres from the bottom of the jet. Do not insulate the flask in anyway, lift the ball valve and simultaneously start the time recorder. Suspend the valve from the cup and the thermometer by means of the hook in the wire stem. Stop the time recorder when the sample reaches the graduation mark of the flask and note the final reading of the oil cup thermometer.

Reject any determination if the temperature of the sample in the oil cup varies during the run by more than 0.1°C for temperature of 60°C or below, by more than 0.3°C for 93°C or below and by more than 0.85°C at 121°C.

Calculation: Report the time in second to the nearest 0.5°C for values below 200 seconds and to the nearest whole second for values above 300 seconds.

Report: Report the Redwood viscosity in seconds along with the viscometer number and test temperature.

Experiment 15

Determination of Percent Extractable Ingredients of a Compounded Rubber by Solvent Extraction method

Principle: To evaluate the concentration of some of the ingredients using Soxhlet extraction is a tedious method. The compounded rubber contains many extractable ingredients such as plasticizers, processing aids, waxes, antioxidants and vulcanizating systems. Acetone extracts some of the above ingredients either as it is or in degraded form. This method is applicable to extract ingredients from crude, unvulcanized, reclaimed or vulcanized NR, SBR, BR and IIR.

Apparatus: The Soxhlet's extraction apparatus consisting of a round-bottom flask, glass extraction unit, water cooled condenser, heating mantle, analytical balance (Fig. 3.11).

Materials: Rubber sample, acetone.

Procedure: The given compounded rubber (around 2–3 g) is weighed with the filter paper. It is cut into very small pieces and wrapped with filter paper, then put into the extraction cup. In the 250 ml round-bottomed flask, around 120–150 ml of acetone is taken. The extraction liquid (acetone) is heated to boiling in a round-bottom flask, and the resulting vapour is condensed in a reflux condenser mounted at the top of the extractor. Acetone condenses on the rubber and gets filled in the extraction cup. Once it reaches the maximum levels in the middle container, it falls back into the round bottom flask.

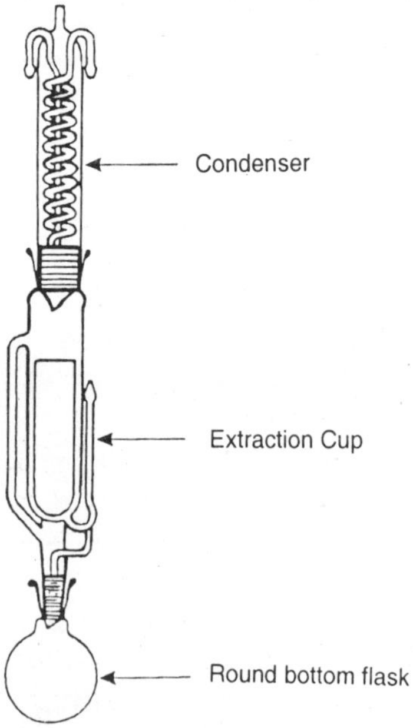

Fig. 3.11: *Soxhlet Extraction Apparatus*

The solvent must have a lower specific gravity than the material being extracted (otherwise the sample would float on the solvent in the extractor cup). Like this about 15–25 cycles of acetone should be passed over the rubber. Finally, the extracted rubber along with the filter paper is dried and weighed.

The percentage acetone extract can be calculated from the following equation:

$$\text{Percentage acetone extract} = \frac{W_2 - W_3}{W_2 - W_1} \times 100$$

where W_1 = weight of the filtar paper, W_2 = weight of the sample before extraction along with the filter paper, and W_3 = weight of the sample after extraction along with the filter paper.

Result: Report the percentage of acetone extract.

Experiment 16

Estimation of Number Average Molecular Weight by End Group Analysis

Principle: The end group analysis is a chemical method used to determine the number average molecular weight of a polymer sample which contains reactive functional groups at the chain ends. Usually, the total number of functional groups present in a given weight of a polymer sample is expressed as a functional equivalent per 100 g. From the knowledge of functionality and functional group equivalent, number average molecular weight, $\overline{M}_n$ can be found out by using the relation:

$$\overline{M}_n = \text{Functionality/Functional group equivalent} \qquad (1)$$

Apparatus: Electronic balance, 250 ml iodine flask, hot plate cum magnetic stirrer, pipette and burette.

Materials: Ethanol, carobxy terminated polybutadience (CTPB), 0.5 N KOH and phenolphthalein.

Procedure: About 0.5 g of sample (CTPB) is taken in a 250 ml iodine flask, 25 ml of ethanol and 25 ml of toluene are added to it. This mixture is refluxed for 25–30 minutes with constant shaking or stirring. Cool the fask, wash the condenser with little alcohol and collect all washings to the same flask. Titrate it against 0.5 N NaOH using phenolphthalein as an indicator. Similarly, carryout the blank titration without sample. The functional group equivalent is calculated by using the relation:

$$\text{Functional group equivalent} = \frac{(B - S) \times N_{NaOH}}{10 \times \text{Weight of sample}} \quad (2)$$

where, B is blank titre value; S is sample titre value; N_{NaOH} is normality of NaOH and W is weight of the sample taken.

The functionality of linear molecule is considered as two, because linear molecules has two ends. $\overline{M}_n$ can be calculated by using equation(1).

Disadvantages: Eventhough end-group analysis is a simple, low cost and feasible technique, but it has the following disadvantages: (i) The polymer should be linear, (ii) It should possess functional group such as —OH, —COOH, —CHO, —NH_2, etc. (iii) It should be soluble in common solvents. Molecular weight up to 30,000 can be measured, beyond that it is not possible by this method.

Result: Report the number average molecular weight of the given polymer sample.

Experiment 17

Estimation of the Extent of Swelling of Polymer in Different Solvents

Apparatus: Electronic balance, four beakers.

Materials: Water, acetic acid, formic acid, hexane and cross linked polymer.

Procedure: About 1 g of the cross-linked polymer (polyacrylamide cross-linked with N, N^1–MBA) is weighed into each of the four beakers. The four samples are dipped in four different solvents in different beakers. It is kept undisturbed for 48 hours at room temperature and then taken out. The solvent is filtered off, the sample is wiped with filter paper and weight is noted.

Calculation:

$$\text{Swelling co-efficient (Q)} = \frac{M - M_o}{M_o} \times \frac{1}{\rho}$$

where, M_o is initial weight of polymer sample, M is weight of the swollen sample and ρ is the density of the solvent.

Result:

$Q(H_2O)$ = ...

$Q(CH_3COOH)$ = ...

$Q(HCOOH)$ = ...

Q(hexane) = ...

Experiment 18

Estimation of Free Acid content, Acetyl Value and Hydroxyl Value of Cellulose Acetate

Introduction: Cellulose is the most abundant constituent of the plant kingdom. Cellulose content is over 90 per cent in cotton and nearly 50 per cent in an average wood. The most convenient sources of cellulose for chemical treatments and transformations are cotton linters and wood pulp.

Of the very large number of cellulose derivatives that have been prepared, only the esters and ethers are generally important. The three-hydroxyl groups of the repeating anhydro-2-glucose units are transformed into ester or ether groups, during esterification and etherification, giving a theoretical degree of substitution (DS) equal to 3.0. Among the important cellulose esters are: cellulose nitrate (CN), cellulose acetate (CA), cellulose acetate butyrate (CAB) and cellulose propionate (CP).

The three stages in the manufacturing of cellulose acetate are: (i) pre-treatment of cellulose, (ii) acetylation, and (iii) hydrolysis. Cellulose plastics are made by substituting various chemical groups for the hydroxy groups for an average 2.4 of the three hydroxyl groups per six carbon units, produce a moulding grade cellulose acetate. Commercially, three kinds of cellulose acetates are available. They are: (i) primary cellulose acetate, (ii) secondary cellulose acetate, (CSA) and (iii) tertiary cellulose acetate (CTA). These experiments can be used to distinguish from one to another types of cellulose acetates and to estimate the degree of acetylation.

The acetyl content is a measure of the degree of esterification of cellulose. Properties of the cellulose acetate, such as solubility and moisture sensitivity, and the dyeing characteristics of the yarn are affected by the acetyl content.

In characterizing the degree of acetylation, "percent acetyl value" and "per cent combined acetic acid" are both used. Since the formula weights (M.W.) are: —CH_3CO, 43, and CH_3COOH, 60, the two figures are always in the ratio 43:60. Their relation to the number of hydroxyls acetylated per glucose residue is as follows:

Acetyls per glucose	*Percent acetyl value*	*Combined acetic acid*
2.0	34.9	48.8
2.5	40.2	56.2
3.0	44.8	62.5

In the procedures that follow, a weighed sample of acetate or triacetate flake or fibre is dissolved in a suitable solvent, precipitated in a finely divided state, and saponified by the addition of a measured excess of alkali. The quantity of alkali used for saponification is determined and calculated as percent acetic acid by weight of dry cellulose acetate or triacetate. The standard deviation (s.d.) of secondary acetate with a maximum percent acetic acid ranging from 45 to 57% is ± 0.06%. for triacetate the values from 57 to 63%, the standard deviation is ± 0.08%.

(a) Estimation of free acetic acid content

Purification of cellulose acetate requires extensive washing to remove the residual acids. Even so, trace levels of acidity exists. The method of analysis is a water extraction and direct titration with dilute sodium hydroxide.

Apparatus: Analytical balance, pipette, burette, 250 ml iodine flask, hot plate cum magnetic stirrer.

Materials: Cellulose acetate, phenolphthalein, 0.01 N NaOH.

Procedure: About 5 g cellulose acetate sample, corrected for moisture content is weighed and ground to pass a No. 20 sieve, then added into a 250 ml flask. 150 ml freshly boiled water is added and cooled to room temperature. The flask is stoppered, shaked and allowed to stand for 3 hours. Then cellulose acetate is filtered off and washed with 25 ml of water in three portions. The combined filtrate and washing is titrated with 0.01 N sodium hydroxide solution, using phenolphthalein as an indicator. A blank titration is also run on the same volume of water used for the sample.

Free acidity is determined by using the relation,

$$\text{Free acetic acid} = \frac{(S - B)\,(N)\,(6.005)}{W}$$

where, S = volume of sodium hydroxide to titrate sample, in ml; B = volume of sodium hydroxide to titrate blank, in ml; N = normality of sodium hydroxide solution; W = weight of oven dried sample, in g and 60 is the equivalent weight of the CH_3COOH (acetic acid) group (6.005 after conversion).

(b) Estimation of acetyl value of cellulose acetate

(i) Cellulose Secondary Acetate

Apparatus: Analytical balance, pipette, burette, 500 ml Erlenmeyer flask, hot-plate-cum magnetic stirrer.

Materials: Cellulose acetate, acetone, phenolphthalein, NaOH, sulphuric acid.

Procedure: Weigh out 2.1 ± 0.005 g of sample into a weighing bottle. If the sample is yarn or fabric, remove any finish and/or size by extraction with ethyl ether followed by extraction with hot water. Air-dry the sample before weighing. Place the weighing bottle, with the cover removed, in an oven and dry for 1 hour at 120–125°C or 2 hour at 105°C. Remove the bottle from the oven, cover it, and place it in a desiccator to cool. Weigh the closed bottle to the nearest milligram and transfer the cellulose acetate to a 500 ml Erlenmeyer flask that has been rinsed with water. Re-weigh the bottle to the nearest milligram to obtain the weight of the sample by difference.

Wash down any cellulose acetate clinging to the neck and walls of the flask with 10 ml of water and add 150 ml of acetone from a graduated cylinder. Prepare a blank of 150 ml of acetone and 10 ml of water and carry through the remaining steps.

Carefully introduce a Teflon-coated stirring bar into each flask, stopper the flask, and place it on a magnetic stirrer until solution is complete, about 30 minutes. Avoid splashing the solution because the splashing will deposit cellulose acetate on the walls of the flask and result in its loss from reaction. Examine each flask to make sure that the particles of cellulose acetate are all in contact with the solvent and that the solution is proceeding in a normal manner. When solution is complete, place the flask on a small magnetic stirrer under a 30 ml burette or pipette, and adjust the magnetic stirrer so that vigorous stirring occurs. Allow 30 ml of 1N sodium hydroxide to flow freely into the solution. It is absolutely necessary that a finely divided precipitate of regenerated cellulose free from lumps be obtained. If the sample is lumpy at this point, it must be discarded.

Stopper the flask and stir for 30 minutes to allow for complete saponification. Add 100 ml of hot water, washing down the sides of the flask, and stir for 15 minutes more. Remove the blank solution from the stirring unit, add 4 or 5 drops of 1% ethanolic phenolphthalein as an indicator and titrate with 1 N sulphuric acid. Titrate rapidly, using the magnetic stirrer, until the end-point is approached, then finish the titration slowly with rapid stirring until the mixture is pale-pink or just colourless. Record the number of millilitres of sulphuric acid used for the titration to the nearest 0.01 ml. Replace the flask on the stirring unit.

Remove the flask containing the sample from the stirring unit and wash down the sides of the flask with about 25 ml of water. Add 4 or 5 drops of phenolphthalein indicator and titrate the excess sodium hydroxide in the same manner. The solution should remain pale-pink or colourless in each case, indicating that all excess sodium hydroxide was consumed in the titration. Calculate the acetyl content as percent acetic acid by using the following relation:

$$\text{Per cent } CH_3COOH = \frac{(B - S)\,(N \text{ of NaOH})\,(0.06005)\,(100)}{\text{Wt. of sample, in g}}$$

where, S = volume of sodium hydroxide consumed for sample titration, in ml; B = volume of sodium hydroxide consumed for blank titration, in ml; N = normality of sodium hydroxide solution; W = weight of oven dried sample, in g and 0.0605 is the milliequivalent (mEq) weight of the CH_3COOH group.

(ii) Cellulose Triacetate

Apparatus: Analytical balance, pipette, burette, iodine flask, hot-plate-cum magnetic stirrer.

Materials: Cellulose acetate, acetone, phenolphthalein, NaOH, sulphuric acid, dimethylsulphoxide.

Procedure: Repeat the weighing procedure described above using 1.9 ± 0.05 g of sample. Then proceed as follows:

Gently rotate the flask by hand to distribute and spread the sample in a thin layer in the bottom of the flask. Add 70 ml of acetone to the flask, washing down the sides, and gently swirl the flask until the particles are completely wetted and evenly dispersed. Stopper the flask and allow to stand undisturbed for 5 minutes.

Prepare a blank of 70 ml of acetone and carry through the remaining steps. Carefully pour down 30 ml of dimethylsulphoxide through sides of the flask from a graduated cylinder, rinsing down any particles clinging to the sides. Carefully insert a Teflon-coated stirring bar, stopper the flask, and place it on a magnetic stirring unit.

Stir the contents of the flask until the sample is completely dissolved (20–30 min) and when solution appears to be complete, add 50 ml of acetone, washing down the sides of the flask in the process, and stir for an additional time of 5 minutes. When the solution is completely prepared, place the flask on the small magnetic stirrer under the 30 ml burette or pipette, and adjust the magnetic stirrer so that vigorous stirring occurs. Allow exactly 30 ml of 1N sodium hydroxide to flow freely into the solution. It is absolutely necessary that a finely divided precipitate of regenerated cellulose free from lumps be obtained. If the sample is lumpy at this point, it must be discarded.

Stopper the flask and stir for 2 hour to allow for complete saponification. Add 100 ml of hot water, washing down the sides of the flask, and stir for 15 minutes more. Remove the blank solution from the stirring unit, add 4 or 5 drops of phenolphthalein indicator, and titrate with 1N sulphuric acid. Titrate rapidly, using the magnetic stirrer, until the end-point is approached, then finish the titration slowly with rapid stirring until the mixture is pale-pink or just colourless, and record the number of millilitres of sulphuric acid used for the titration to the nearest 0.01 ml. Replace the flask on the magnetic stirrer and stir for at least 10 minutes.

Remove the flask containing the sample from the stirring unit and wash down the sides of the flask with about 25 ml of water. Add 4 or 5 drops of phenolphthalein indicator and titrate the excess sodium hydroxide. Replace the flask on the magnetic stirrer and stir for at least 10 minutes. Each solution should remain pale-pink or colourless, indicating that all excess sodium hydroxide was consumed in the titration. Calculate acetyl content as described for secondary acetate.

(c) Estimation of hydroxyl value of cellulose acetate

Apparatus: 250 ml iodine flask, microburette, hot-plate-cum magnetic stirrer.

Materials: Dried cellulose acetate, acetylating agent, 0.5N NaOH solution, pyridine, phenolphthalein.

Procedure: Weigh about 2 g of dried cellulose acetate into a 250 ml of iodine flask. Add 5 ml of acetylating agent from a microburette. Prepare a blank of 5 ml of acetylating agent with 25 ml of water and carry throughout the remaining steps.

Introduce magnetic bit into each flask, stoppered the iodine flask with pyridine moistened stopper and place it on a hot-plate-cum magnetic stirrer and reflux it for about 30 minutes. Add 25 ml of water into each flask. Stopper the flask and stir for 5 minutes more, to allow for complete hydrolysis. Add phenolphthalein indicator and titrate with 0.5N NaOH. Titrate immediately, using magnetic stirrer until the mixture is pale-pink. Hydroxyl value is calculated using the relation;

$$\text{Per cent Hydroxyl value} = \frac{(B - S) \times N_{\text{NaOH}} \times 17 \times 100}{\text{Weight of the sample} \times 1000}$$

where, B is the volume of NaOH consumed in ml for blank titration; S is the volume of NaOH consumed in ml for sample titration, N_{NaOH} is the normality of NaOH, and 17 is the equivalent weight of hydroxyl group.

Result: Report the hydroxyl value of the cellulose acetate sample.

Experiment 19

Determination of Plasticizer in Polyvinyl Chloride (PVC)

Introduction: The additives such as plasticizers, heat stabilizers, processing aids, etc., are widely used in plastic industries. Identification of additives in the polymer is one of the most important areas, for the environmental technologists to know their toxic behaviour. Migration of additives from

Table 3.7: *Some examples of commercial plasticizers for plastic materials*

Plasticizers	*Polymer*	*Plasticizer type*
Dioctyl phthalate (DOP)	PVC and copolymers	General purpose, primary plasticizers
Tricresyl phosphate (TCP)	PVC and copolymers	Flame retardant and primary plasticizers
Tricresyl phosphate (TCP)	Cellulose nitrate	Flame retardant and primary plasticizers
Dioctyl adipate (DOA)	PVC, cellulose acetate butyrate (CAB)	Low temperature plasticizer
Dioctyl sebacate (DOS)	PVC	Non-migratory secondary plasticizer
Chlorinated paraffins	Most polymers	Flame retardant, plasticizer, extenders
Bi and tertiary phenyls	Aromatic polyesters	various
Sulphanamide formaldehyde resin	Polyamides	Non-migratory secondary plasticizer

plastic is also equally important to the scientists, because of their malfunctioning or toxicity. These are the important areas one should be aware. Without the addition of plasticizer, PVC cannot be processed into useful packaging material for food packaging. The palsticizes, which are used in PVC, are phthalates. These plasticizers have a potential to migrate into packed material from the packaging and become adulterant.

Apparatus: Soxhlet extraction apparatus, sintered glass crucible, hot air oven, analytical balance.

Materials: Sample (PVC film), ether, 0.5N alcoholic KOH solution, 0.5N HCl, phenolphthalein, anhydrous $CaSO_4$ and ethylene glycol.

Procedure: About 20 g of PVC sample is shredded and packed loosely in a filter paper and placed into a Soxhlet extraction apparatus. About 2C0 ml of ether is taken in the round bottom flask. The sample is extracted with ether for 3–6 hours. The ether extract in the 500 ml flask is placed under suction to evaporate ether. Then the liquid residue is dried in a hot air oven to constant weight. This is the total extractable plasticizer. The residue in the Soxhlet apparatus should also be dried to constant weight, and the loss in weight should be compared with the weight of plasticizer recovered.

The saponification value is determined on a 2 g portion of the plasticizer by refluxing for 2 hour with 50 ml of 0.5N alcoholic KOH. After cooling and diluting with water, the excess of KOH is titrated with 0.5 N HCl using phenolphthalein as an indicator. A control (blank) test is carried out simultaneously on 50 ml 0.5N alcoholic KOH. And the difference in titre value is used to calculate the saponification value (number of mg of KOH consumed per gram of plasticizer).

The phthalate plasticizer can be positively identified by the following procedure. Ethylene glycol (5 ml) is refluxed in a 25 ml pear-shaped micro-casein flask till all material, volatile is below 190 °C removed. Then, 2 ml of the extracted plasticizer and 2 g of KOH are added, and distillation is continued, care being taken to avoid spurting during the hydrolysis, until 1–2 ml of distillate have been collected. This now contains the alcoholic component which may be identified by separating the upper layer, drying with anhydrous $CaSO_4$, and measuring the refractive index at 20°C and the boiling point by the capillary tube method (Siwoloboff).

Result: The report should include:

(a) Extracted plasticizer as per cent of weight of PVC sheet.
(b) Difference in weight of PVC before and after extraction
(c) Saponification value of plasticizers.

Experiment 20

Determination of Network Parameter of Vulcanized Rubber

To determine the average molecular weight between junction points in a cross-linked network by elastic modulus and swelling measurements.

Apparatus: Fine-mesh wire screen, retort stand, 50 cm ruler, bulldog clips and lightweight pan.

Materials: Natural rubber, benzene, phenyl-β-naphthylamine (PBN) and methanol.

Procedure: The freshly unpacked rubber band is washed, dried with tissue or filter paper and cut into two strips, each 7 cm long. The density of the rubber (g/cc) is determined by flotation on pieces of about 5 mm long. Twenty-five ml water is taken in 100 ml beaker, upon which the rubber floats. Methanol is run in to the water, from a burette, with occasional agitation, until the pieces of rubber neither float nor sink. The density of the rubber now just matches that of the aqueous solution, which may be determined by means of a specific gravity bottle.

Stress–Strain curve: The bulldog clips and pan are weighed. Two bulldog clips are affixed to the rubber band exactly 5 cm apart. The assembly is suspended vertically, and the distance between the clips is measured with 50 cm rule. The pan is hung on, and the measurement is repeated. Weights are now added progressively, and separation is measured until it reaches 25 cm. The weights are now progressively removed, and again the separation is recorded for each weight. It is important to ensure there is no slippage at the clips. If this happens, either stronger clip must be used or the rubber strips must be cut longer to allow looping in the clip.

If α represents the ratio of elongated length to initial length, the extension $x = \alpha - 1$. The total load in grams, i.e. weights plus pan plus lower bulldog clip, should now be plotted firstly against $(\alpha - 1)$ and then against $[\alpha - (1/\alpha^2)]$. In the second case, a straight line should pass through the points at least up to 300 % extension. The slope of this line should be recorded. The unit will be in grams. Using this slope (S), the molecular weight between junction points M_c can be calculated from:

$$M_c = \frac{WRT}{981\,S}$$

where, W is the weight per cm length of the rubber band; T is the absolute temperature; R (gas constant) = 8.34×10^7 ergs/mole degree.

Swelling measurement: The other rubber strip is accurately weighed and added into a conical flask containing 0.5 g phenyl–β–naphthylamine (PBN) (or similar antioxidant) and 50 ml of distilled benzene, which is then stoppered and kept in dark for 1 week. After this time, the benzene is poured off, the gel is held for 5 seconds on a fine-mesh wire screen, then dropped into a weighing bottle which is covered and weighed. The volume fraction V_2 of the swollen rubber in the gel can be calculated from the weight of rubber gel W_g. The initial weight of the rubber specimen W, the density of the rubber d, and the density of benzene D, since the volume ratio Q_m of solvent to rubber in the gel is given by:

$$Q_m = \left(\frac{W_g - W}{D} \middle/ \frac{W}{d}\right)$$

and $$V_2 = 1/1 + Q_m$$

Given that the molecular weight of benzene is 78.11 and its density (at 25°C) is 0.8794 g/cc, its molar volume (V_0) is $\frac{78.11}{0.8794} = 89.8\ \text{cm}^3/\text{mole}$. Given also that the interaction constant for natural rubber/benzene, $\mu = 0.43$, the molecular weight between cross-links may be calculated from:

$$Mc = \frac{-dV_0V_2^{1/2}}{\text{In}(1 - V_2) + V_2 + \mu V_2^2} = \frac{-89.8\ dV_2^{1/2}}{\text{In}(1 - V_2) + V_2 + 0.43V_2^2}$$

A less accurate expression is sometimes used

$$M_c = \frac{dV_o}{1/2 - \mu} V_2^{-5/3} = 1283\ dV_2^{-5/3}$$

This equation implies that the volume fraction is proportional to the 0.6 power of the cross-link density.

Result: Report the density of rubber and the value of molecular weight between cross-links of the given polymer sample.

Experiment 21

Determination of Nitrogen in a Polymer using Micro Kjeldahl apparatus (Polymer sample—NBR or polyacrylamide)

In this method a polymer solution is treated with a solution of a strong base (e.g., sodium hydroxide) and the mixture is distilled. Nitrogen (in the form of ammonia) is quantitatively expelled and is absorbed in an excess of standard acid. The excess of acid is back titrated in the presence of methyl red or methyl orange indicator. Each ml of acid consumed in the reaction is equivalent to 0.014 g of nitrogen.

Apparatus: 250 ml beaker, Kjeldahl flask (Fig. 3.12), reflux condenser, burette, pipette, analytical balance, conical flask.

Materials: Sample (polyacrylamide), suitable solvent, 0.1M HCl, 10% NaOH solution, 0.1 M NaOH methyl red indicator.

Procedure: Weigh out accurately about 2 g of polymer (polyacrylamide) sample into a 250 ml beaker. Dissolve it in a suitable solvent (water). Transfer the above solution into the Kjeldahl flask and dilute with 200 ml of water (solvent). Place 100 ml of standard 0.1M hydrochloric acid in the receiver and adjust the flask so that the end of the condenser completely dips into the acid. Make sure that all the joints are fitting tightly. Place 100 ml of 10% sodium hydroxide solution in the separating funnel. Run the sodium hydroxide solution into the Kjeldahl flask by opening the tap; close the tap as soon as the

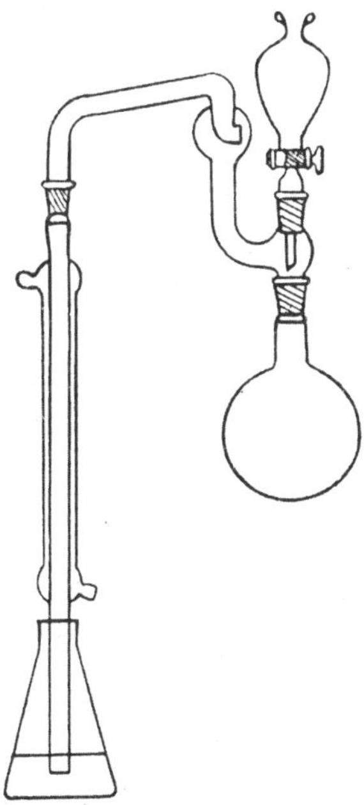

Fig. 3.12. *Ammonia distillation (Kjeldahl) apparatus for nitrogen estimation*

alkali has entered. Heat the flask so that the contents boil gently. Continue the distillation for 4–5 minutes, by which time all the nitrogen should have passed over into the receiver; open the tap before removing the flame. Disconnect the tap from the top of the condenser. Lower the receiver and rinse the condenser with little solvent. Add a few drops of methyl red indicator and titrate the excess acid in the solution with standard 0.1M sodium hydroxide. Repeat the experiment in duplicate and calculate the percentage of nitrogen in the given polymer sample.

1 ml 0.1M HCl = 1.4 mg nitrogen

Result: Report the nitrogen content of the given sample.

Experiment 22

Determination of Chlorine Content Present in the given Polymer by Gravimetric Method

The aqueous solution of the chloride is acidified with dilute nitric acid in order to prevent the precipitation of other silver salts, such as the phosphate and carbonate, which might form in neutral solution, and also to produce a more readily filterable precipitate. A slight excess of silver nitrate solution is added. Silver chloride is precipitated.

$$Cl^- + Ag^+ \rightarrow AgCl \text{ (ppt.)}$$

The precipitate, which is initially colloid in nature, is converted into a curd-like precipitate by heating the solution and stirring the suspension vigorously. The precipitate is collected in a filtering crucible, washed with very dilute nitric acid (in order to avoid becoming colloidal), dried at 120–130°C, and weighed as AgCl.

The solubility of AgCl in water at 20°C is 1.4 mg/lt and at 100°C it is 21.7 mg/lt. The solubility is less in the presence of ammonium salts or alkali metal salts, and in large concentrations of acids. If silver chloride is washed with pure water, it may become colloidal and run through the filter. Nitric acid is generally employed because it has no action on the precipitate and is readily volatile (concentration of nitric acid should not exceed 0.01M). Completion of washing of the precipitate is tested when chloride ions have been removed.

This may be done by adding 1–2 drops of silver nitrate to the washings collected, which is acidified with dilute nitric acid. Silver chloride is sensitive to light, decomposition occurs into silver and chloride. The silver remains colloidally dispersed in the silver chloride and thereby imparting a purple colour to it. The decomposition by light is superficial, and is negligible when the precipitate is exposed to direct sunlight and stirred frequently. Hence, experiment must be carried out in a dark place. When the solution containing the precipitate is set aside, it should be placed in the dark or the vessel containing silver chloride should be covered with a black or a used carbon paper.

Apparatus: 400 ml beaker, sintered glass crucible, analytical balance, filter flask.

Materials: Sample, solvent, conc. HNO_3, $AgNO_3$, dil HNO_3.

Procedure: Weigh out accurately about 1 g of the given polymer sample (~0.1 g of chlorine) into a 400 ml beaker containing a glass rod. Add about 100 ml of solvent, stir until the polymer sample has dissolved and add 0.5 ml of concentrated nitric acid. To the cold solution, add 20 ml of 0.1M silver nitrate solution slowly and with constant stirring, only slight excess should be added. Allow the precipitate to settle. Test the supernatant liquid with a few drops of $AgNO_3$ to be sure that precipitation is complete. Do not expose the solution to bright light. Heat the solution nearly to boiling, while stirring constantly, and maintain it at this temperature until the precipitate coagulates and the supernatant liquid is clear (2–3 min). Again check the supernatant liquid for chlorine. Allow the precipitate to settle in a dark place for about an hour. Filter through a previously weighed sintered glass crucible by transferring the supernatant liquid initially. Wash the precipitate 2 or 3 times with very dilute nitric acid (~0.1M), and wash until the precipitate is free from free silver ions. Dry the precipitate at 120–130°C for 1 hour, allow to cool in a desiccator and weigh. Repeat the heating and cooling until constant weight is obtained.

One M.Wt. (143.4 g) of AgCl contains = 35.5 g of Cl^-

'a' g of AgCl contains = $a \times 35.5/143.4$ g of Cl^-

Result: Report the chlorine content present in the given sample.

Experiment 23

Estimation of Periodic Acid induced Degradation of Polyvinyl Alcohol

Purpose: To illustrate a chemical method of degradation with a specific reagent and to ascertain the proportion of head-to-head links in polyvinyl alcohol (PVA).

Apparatus: Dilution viscometer with a flow time of 50–70 s for water.

Reaction:

$$\text{—}CH_2\text{—}\underset{\displaystyle OH}{\underset{|}{CH}}\text{—} \xrightarrow{HIO_4} 2\text{—}CHO + H_2O$$

Materials: Polyvinyl alcohol, periodic acid.

Procedure: An aqueous solution of polyvinyl alcohol is prepared by weighing 2 g into a 100 ml volumetric flask, adding 80 ml distilled water, heating in a beaker of boiling water till dissolved, cooling and then supporting in a thermostatic bath at 25°C and adding sufficient water to makeup to the mark. A solution of periodic acid (1.7 g) or of sodium periodate (1.9 g) in distilled water is prepared and made upto 50 ml volumetric flask held in the thermostat. This solution is 0.45M. Pipette out 5 ml filtered water and 5 ml filtered periodic acid solution into the dried and clean dilution viscometer. These are mixed and mean flow time is measured. Let this flow time be known as t_a.

The viscometer is now cleaned and dried, then 5 ml of polyvinyl alcohol solution and 5 ml of water are pipetted in. After mixing the mean flow time is recorded. Let this flow time be known as t_b.

The viscometer is once again cleaned and dried then 5 ml PVA solution and 5 ml periodic acid solution are added. The solutions are mixed and the flow time measured within 5 minutes of the mixing. Then, the flow time is measured repeatedly at 5 minutes interval until it becomes constant. Let the final value be known as t_c.

Calculations: Assuming that the viscosity contribution of the periodic acid is negligible, the relative flow time of the non-degraded polymer is t_b/t_a. The limiting viscosity number of the non-degraded polymer may be calculated from the approximate relationship:

$$[\eta]_0 = \ln (t_b/t_a)/(c/2)$$

where, c is the concentration of the polymer solution g/100 ml. The relative flow time of the degraded polymer is t_c/t_a. The limiting viscosity number $[\eta]$ of the degraded polymer may be similarly calculated from t_c/t_a, using the same value of c.

The viscosity-average molecular weight $\overline{M}_v{}^0$ and $\overline{M}_v$ of initial and degraded polymer may now be calculated from $[\eta]_0$ and $[\eta]$, using the relationships:

$$[\eta]_0 = K(\overline{M}_v{}^0)^\alpha$$

$$[\eta] = K(\overline{M}_v)^\alpha$$

where, $K = 2 \times 10^{-4}$ and $\alpha = 0.76$

The number-average molecular weights may be calculated using the approximate relation for a random polymer:

$$\overline{M}_v = 1.85\,\overline{M}_n$$

If the decrease in molecular weight is due only to cleavage of the polymer chains at sites where 1, 2 –glycols are situated, then the weight fraction of 1, 2 –glycols is:

$$x = 44\,\{(1/\overline{M}_n) - (1/\overline{M}_n{}^0)\}$$

Result: The report should include the viscosity-average molecular weights of the initial and degraded polyvinyl alcohol, the time required for the degradation to be complete and mole per cent of the 1.2 glycol units.

Experiment 24

Determination of Miscibility of Polymer Blends

Introduction: There are various techniques to study the miscibility of the polymer blends. X-ray diffraction, differential scanning calorimetry (DSC), scanning electron microscopy (SEM), infrared spectroscopy, etc., are some of the techniques which are widely used. But these techniques are time consuming, expensive, and complicated. At the same time these techniques cannot be used for routine analysis in an industry for the reasons mentioned above. In order to study the miscibility of the polymer blends, simple, rapid and low-cost techniques such as, (i) intrinsic viscosity method, (ii) refractive index method, (iii) density measurement method, (iv) ultrasonic velocity measurement method, and (v) optical polarizing microscopy are widely used.

(a) Miscibility of Polymer Blends by Viscosity Method

The relative viscosity of the polymer blend is determined using an Ubbelhode Suspended Level Viscometer (USLV). It is a simple capillary device. The advantage of using USLV is that the measurements are unaffected by the volume of the solution taken. As such the concentration of the solution taken can be varied within the viscometer by adding appropriate quantity of the solvent, and the corresponding time of flow can be measured. As the viscosity is highly dependent on the temperature, all the measurements must be carried out at a constant temperature ($\pm$ 0.01°C) by immersing the viscometer (up to neck level) in a constant temperature water bath fitted with glass window.

Apparatus: Ubbelhode Suspended Level Viscometer (USLV), constant temperature water bath, thermometer and standard flask.

Materials: Polymer, solvent.

Procedure: Prepare solutions of the polymers considered for miscibility studies in a suitable common solvent. Maintain a total polymer concentration of 0.5–1 g/dl. Place 10 ml of the prepared solution into USLV and determine the flow time t_c. Dilute the solution successively with a known amount of the solvent and measure flow time for those concentrations. Record the time of flow for the solvent (t_0) alone and calculate the reduced viscosity $[(t_c - t_0)/t_0]$. Repeat the steps for the constituent polymers. Draw the plots of reduced viscosity versus concentration for the blend as well as the constituent polymers.

The Huggin's equation which relates specific viscosity, η_{sp}, with intrinsic viscosity, $[\eta]$ and concentration, C of single solute solution is given by:

$$\eta_{sp}/C = [\eta] + bC$$

and the relation between constant b and Huggin's coefficient, K_H is given by:

$$b = K_H [\eta]^2$$

For ternary systems containing a solvent and two polymers, Chee suggests:

$$b = W_1^2 b_{11} + W_2^2 b_{22} + 2W_1 W_2 b_{12}$$

An arithmetic differential parameter defined by the following equation is a simple measure of intermolecular interaction perameter when the polymers are mixed in weight fractions W_1 and W_2 as;

$$\Delta B = (b - \bar{b})/2W_1 W_2$$

and

$$\bar{b} = W_1 b_{11} + W_2 b_{22}$$

Chee defined a more effective parameter, μ to predict polymer miscibility given by the equation:

$$\mu = \Delta B/\{[\eta]_2 - [\eta]_1\}^2$$

Sun *et al* suggested a new equation for the determination of miscibility of the polymer, considering the long-range hydrodynamic interaction of pairs of single molecules.

$$\alpha = K_m - \frac{\left(K_1[\eta]_1^2 W_1^2 + K_2[\eta]_2^2 W_2^2 + \sqrt{2(K_1 K_2)}\,[\eta]_1[\eta]_2 W_1 W_2\right)}{\left\{[\eta]_1 W_1 + [\eta]_2 W_2\right\}^2}$$

where, W_1 and W_2 are the weight fractions of the two polymer components. b_{11} and b_{22} are the slopes of the viscosity curves for the two polymer components and b_{12} is the slope of the viscosity curve for the blend solution. $[\eta]_1$ and $[\eta]_2$ are the intrinsic viscosities for the pure component solutions. K_1, K_2 and K_m are the Huggin's constants for individual components 1, 2 and blend respectively. The blend would be miscible if $\Delta B \geq 0$, $\mu \geq 0$ and $\alpha \geq 0$ and immiscible when $\Delta B < 0$, $\mu < 0$ and $\alpha < 0$.

(b) Miscibility of Polymer Blend by Density

The density of the polymer blend solution is measured using a specific gravity bottle. The density is given by the relation:

$$\left.\begin{array}{c}\text{Density of}\\ \text{polymer blend}\end{array}\right\} = \frac{\left\{\begin{array}{c}\text{Wt. of the polymer}\\ \text{blend solution}\end{array}\right\}}{\left\{\begin{array}{c}\text{Wt. of water at}\\ 30^\circ\text{C}\end{array}\right\}} \times \left\{\begin{array}{c}\text{Density of}\\ \text{water at}\\ 30^\circ\text{C}\end{array}\right\}$$

Prepare the polymer solution and their blend solutions at different compositions maintaining a concentration of 1g/dL. Note the weight of a clean, dry specific gravity bottle. Take water in the specific gravity bottle and record the weight. Wipe the specific gravity bottle with a tissue paper to remove any traces of water and record its weight. Repeat the procedure for all the blend compositions prepared. Calculate the densities for all the compositions of blends. Plot a graph of density versus composition. The straight line indicates the miscibility of the blend.

(c) Miscibility of Polymer Blend by Refractive Index Method

Refractive index (η_D) of the polymer blend solutions is measured using an Abbe's refractometer.

Apparatus: Abbe's refractometer and mercury light source (Fig. 3.13).

Materials: Polymer blend solutions of different compositions.

Procedure: Prepare the polymer blend solutions at different compositions maintaining a concentration of 1g/dL. Maintain

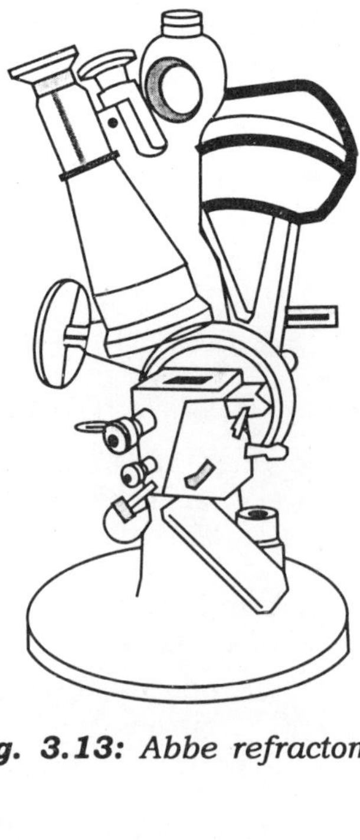

Fig. 3.13: *Abbe refractometer*

the temperature of the sample cell/prism in the refractometer constant by circulating water at the required level using a thermostat. Use sodium vapour lamp (λ = 5893 Å) as the source of the light. Before measuring the refractive indices of the blend solutions, calibrate the refractometer with double distilled water and some simple organic solvents. The refractive index is directly read by moving the index arm of the refractometer until the field seen through the eyepiece is one-half-dark. Adjust the compensator arm to remove all colours from the field. The index arm is then adjusted by means of a Vernier until the dividing line between the light and dark portion of the field exactly coincides with the intersection of the cross wire as seen in the eyepiece. In order to make the dividing line sharp, place a tracing paper over the reflecting mirror. This provides a better source of diffused light. Clean the prisms with soft tissue paper moistened first with the solvent and then with acetone before every use. Note the refractive indices of all the blend compositions and plot a graph of refractive index verses composition. The straight line indicates the miscibility of the blend.

(d) Miscibility of Polymer Blend by Ultrasonic Velocity Measurements

Ultrasonic velocity measurements of the polymer blend solutions of different compositions can be carried out by an interferometric technique. The constant temperature of the instrument is maintained by circulating cold water.

Principle: The principle used in the measurement of sound velocity (ν) is based on the accurate determination of the wavelength (λ) in the medium. Ultrasonic waves of known frequency are produced by a quartz crystal fixed at the bottom of cell. When the waves are reflected by a movable metallic plate in exact multiples of this wavelength, standing waves are formed in the medium. The acoustic resonance gives rise to an electrical radiation on the generator driving the quartz plate and the anode-current of the generator attains a maximum.

If the distance is now increased or decreased with the help of the micrometer screw, and the variation is exactly one-half of length (1/2) or multiple of it, the anode current again becomes maximum. The following relation can be used to calculate the velocity when the wavelength is known:

$$\text{Velocity} = \text{Wavelength} \times \text{frequency}$$

Apparatus: Interferometer, thermostat, thermometer

Materials: Polymer samples, solvents.

Procedure: The measuring cell is connected to the output terminal of a high frequency generator. The cell is filled with experimental solution before switching on the generator. The temperature of the solution inside the cell is maintained at the desired value by circulating the water from the thermostat maintained at that temperature. The measurements should be made only after equilibrium is achieved. Generally, it takes at least 30 minutes for this. The ultrasonic waves move normal from the crystal till they are reflected back from the movable plate and standing waves are formed in the samples in between the reflector plate and the quartz crystal.

The micrometer is slowly moved till the anode-current meter on the high-frequency generator shows a maximum. A number of maxima readings of the anode-current are obtained by moving the micrometer and their number, n, is counted. The total distance (D), thus, moved by the micrometer can be calculated with the help of the following relation:

$$D = n + \lambda/2$$

Once the wavelength is known, the velocity (v) in the sample can be calculated. Each time the cell is thoroughly rinsed with solvent followed by acetone. The cell is then dried with clean hot air. For each composition of the blend the experiment is repeated 10 times and the average velocity is reported.

Experiment 25

Estimation of the Resistance of Plastic Films to Chemicals and to Measure the Weight Change of Films after Immersion in Chemicals as per ASTM D 1239–55 method.

Significance: The highest consumption area of plastic is in packaging. This is mainly due to the superiority of plastics over other materials in terms of inertness and stability to chemical environments like, water, oxidizing agents, oils, lubricants, industrial solvents and foodstuffs. Similarly, chemical resistance is one of the reasons why plastics are found suitable for other industrial applications like self-lubricating joints in chemical units, pipes, tubing and fuel tanks for automotives and indicators for oil and lubricants. The selection of particular plastic materials for specific industrial applications has to be based on the chemical environment to which they will be exposed. There are chemical substances that are hostile and others are benign to plastics.

Apparatus: Beakers, electronic balance, hot air oven.

Materials: Clear plastic sheets like PVC, benzene, sodium chloride (10%), ethanol, ammonia, conc. HCl, acetic acid (10%), distilled water.

Test specimens: The test specimens of the plastic sheets should be in the form of square of 50 mm. Atleast two specimens of each sample should be tested with each chemical reagent.

Procedure: The test specimens should be conditioned at 25 ± 1°C for sometime in the standard laboratory atmosphere. Also the chemical reagents should be maintained at the test temperature for atleast 4 hours before the specimens are immersed in it. The specimens are weighed and immersed in the beakers containing 250 ml of the reagents. The specimen is suspended freely in a vertical position, but fully covered by the reagent. The beakers are kept intact for 24 hours. The specimens are removed and wiped gently with a soft dry cloth or with the filter paper. Specimens taken from water or volatile solvent like benzene require no rinsing, but wiped dry as said before. Specimens tested in salt solutions, acids or alkalis are rinsed with water before wiping to dryness. The percentage of loss or gain in weight from extraction expressed as percentage weight loss or gain compared to the original specimen weight is calculated as follows:

Percentage of weight loss or gain = $(W_1 \sim W_2) \times 100/W_1$

where, W_1 and W_2 are the weight of the specimen before and after exposure to different chemical reagents respectively.

Result: Report the weight loss or gain in percentage of the sample and the effect of each chemical reagent.

Experiment 26

Determination of Free Sulphur and Total Sulphur from Vulcanized Rubber

ASTM D 297–93 is one of the methods to determine the free and total sulphur from the vulcanized rubber.

(i) Free Sulphur

Scope: The sulphur in a rubber compound available for further vulcanization or excess sulphur is extractable by sodium sulphite. This sulphur is determined by titrating the sodium thiosulphate resulting from the extraction. It represents essentially elemental sulphur but will include some of the polysulphide sulphur especially in the presence of thiuranium disulphide compounds.

Apparatus: Buchner funnel, analytical balance, hot air oven.

Materials: Cadmium acetate solution, 40 % formaldehyde solution, glacial acetic acid, 0.05 N iodine solution, sodium stearate suspension in water, 10 % sodium sulphite solution, 1 % starch solution, and 2 % strontium chloride solution.

Procedure: Cut about 2 g of vulcanized rubber sample into smaller pieces and place in a 250 ml conical flask. Add 100 ml of 10 % sodium sulphite solution, 5 ml of sodium stearate suspension and 1 g of paraffin. Cover the flask with a small watch glass and gently boil for 4 hours. Add 100 ml of 2 % strontium chloride solution and 10 ml of 15 % cadmium acetate solution. Separate the rubber and the precipitate by filtration, using Buchner funnel with suction. Wash twice with cadmium acetate solution. To the filtrate add, 10 ml of formaldehyde solution with stirring, 10 ml glacial acetic acid and 5 ml of 1 % starch solution. Add enough crushed ice to bring down the temperature below 15°C and titrate with 0.05N iodine solution using starch as an indicator near the end-point until get blue colour. Run a blank titration on the reagents. Calculate the percentage of free sulphur as follows:

$$\text{Free sulphur, Percentage} = \frac{(A - B) \times N_{I_2} \times 0.032}{W \times 100}$$

where, A is volume of iodine solution required for titration of the sample; B is volume of iodine solution required for titration of the blank; N_{I_2} is normality of the iodine solution, and W is weight of the sample used.

(ii) Total Sulphur

Scope: All the sulphur present in a rubber compound shall be determined by the Fusion method.

Summary: The sample, wrapped in a filter paper is burned in an oxygen combustion flask; the carbon and hydrogen of the organic matter are oxidized and the sulphur is converted to sulphate ions by combustion and absorption in hydrogen peroxide. The sulphite is titrated with barium perchlorate using thorin as an indicator solution.

Apparatus: Oxygen combustion flask, infrared safety igniter, microburette, magnetic stirrer and beaker.

Materials: Barium perchlorate (0.01N), chlorophosphonazo III indicator, 2M HCl, 1% hydrogen peroxide, isopropanol, oxygen cylinder, thorin indicator.

Procedure: Place 100 ml of 1% hydrogen peroxide and the stirring bar in the empty oxygen combustion flask. Fold a filter paper twice to form a U-shaped boat. The sample mass should be 40–80 mg for a sulphur level range of 1.4–1.9 %. Place the sample in the boat and weigh to nearest 0.1 mg. Place the boat firmly in the platinum sample carrier hung on the hook of the stopper with the pointed end of the paper projecting outward. Insert a tube from the oxygen cylinder to the bottom of the flask and blow for 0.5 minute. Smoothly remove oxygen tube and close the stopper with a pinch clamp. Place the flask in the infrared igniter chamber. Close the door and turn on the infrared light until the paper ignites. After ignition, stir the sealed flask vigorously on a magnetic stirrer. Remove the pinch clamp, tilt the stopper to release the vacuum and open the flask. Pour the contents of the flask into a 250 ml beaker, wash the flask with water 3 times and collect the washings into the beaker. Add 100 ml of isopropanol, 2 drops of thorin solution and the stirring bar to the beaker, titrate with barium perchlorate drop wise from the burette until the yellow colour changes to a pink colour which should remain for at least 30 seconds. Obtain a blank titration value by carrying out the entire procedure without using a rubber sample. Calculate the total sulphur as given below:

$$\text{Total sulphur, \%} = [(A - B)\ M \times 3.21]/W$$

where, A is barium perchlorate solution required for titration of the sample; B is barium perchlorate solution for titration of the blank; M is molarity of barium perchlorate and W is mass of the sample used.

Result: Report the free sulphur and total sulphur content of the sample.

Experiment 27

Testing of Natural Rubber Latex

Introduction: Latex is obtained from the bark of the rubber tree by tapping. It is a white or slightly yellowish opaque liquid with a specific gravity, which varies between 0.974 and 0.986. Also it contains a variety of non-rubber constituents both organic and inorganic, in addition to rubber. But in general the composition of latex is as follows:

Rubber	30.0 – 40.0%
Protein	2.0 – 2.5%
Ash	0.7 – 0.9%
Resin	1.0 – 2.0%
Sugar	1.0 – 1.5%
Water	55.0 – 65.0%

Fresh latex as it comes out from the tree is slightly alkaline or neutral. It becomes acidic rapidly due to bacterial action. The formation of organic acids neutralizes the negative charge on rubber particles and the latex gradually gets coagulated on keeping. Therefore, fresh latex cannot be kept for long without coagulation. Intensive tapping reduces the rubber content in latex. Thus, latex from daily tapped trees or slaughter tapped trees may have a rubber content as low as 20–25 per cent.

A. Determination of Total Solid Content (% TSC)

Principle: Total solid content is measured as per the ASTM D 1076–88.8. Total solid content is a measure of the non-volatile, non-rubber constituents and rubber hydrocarbons. It is defined as the percentage by weight of the latex which is non-volatile at a distinct temperature in an open atmosphere. TSC can be used to check the presence of the considerable amount of added adulterant, which are non-volatile, but water-soluble.

Apparatus: Petridish, analytical balance, hot air oven and desiccator.

Materials: Sample

Procedure: Weigh out about 2.5 g of the latex and pour it onto the dried and weighed petridish, which may be facilitated by carefully adding 1 ml of distilled water to the latex and gently swirling the dish. Dry the latex specimen in a hot air oven at $100 \pm 2°C$ for 3– hours. Cool the petridish in a desiccator and note down the weight along with the dried latex sample. Repeat drying and weighing till the weight is constant. Run a duplicate test and an average of two determinations shall be taken as a result. Calculate the percentage of total solids as follows:

$$\text{Total solid content} = \frac{\text{Weight of the dried film}}{\text{Weight of the latex taken}} \times 100$$

Result: Report the total solid content of the latex in (%).

B. Determination of Dry Rubber Content (% DRC)

Dry rubber content is determined as per ASTM D 1076–88.9 method.

Apparatus: Porcelain evaporating dish, electronic balance, steam bath, hot air oven.

Materials: Rubber latex, acetic acid.

Procedure: Weigh out approximately 10 g of the sample into porcelain evaporating dish and add distilled water until the total solid content (TSC) is approximately 25%. Add sufficient amount of 2% acetic acid, while stirring over a period of 5 min, to coagulate the latex completely.

Place the porcelain dish on a steam bath for 15–30 min without disturbance. A clear serum should result. Pickup the coagulated latex particles with the main body of the coagulum. Wash coagulum in running water and pass between rolls. Repeat this process 5 times and reduce the sheet thickness to 2 mm and dry at $70 \pm 2°C$ in a hot air oven. Cool in a desiccator and weigh. Repeat drying and weighing process until the concordant mass value. Duplicate samples shall be run and report the average. Calculate the dry rubber content as follows:

$$\text{Dry Rubber Content, (\%)} = \frac{\text{Weight of dry coagulum film}}{\text{Weight of the sample}} \times 100$$

Result: Report the dry rubber content in (%).

C. Determination of Mechanical Stability

Scope: The mechanical stability of the latex is calculated as per ASTM D 1076–88. This method covers the determination of the mechanical stability of concentrated natural rubber latex using a high-speed stirring technique (14,000 rpm).

Apparatus: Mechanical stability tester, test bottle.

Materials: Rubber latex, aqueous ammonia.

Latex positioning: The temperature of the sample should not be lowered artificially and the test should be carried out within 24 hour of opening. The latex sample should not be exposed to air.

Procedure: Dilute 100 g of latex sample with aqueous ammonia solution to exactly 55.0 ± 0.2 % total solid contents and without delay warm the diluted latex with gentle stirring to 36–37°C. Immediately filter through a 180 microns stainless steel sieve and weigh 80 g of strained latex into the test bottle. Place the test bottle in position for stirring such that the bottom of the agitator disk is 12.7 mm from the bottom of the bottle. Stir the latex at 14,000 ± 200 rpm until the end-point is reached. The end-point is determined by frequently dipping a clean glass rod into the latex to about 15 mm and drawing it once gently over the palm of the hand. The first appearance of the small particles of coagulated rubber so deposited is the end-point of the test, which should be confirmed by the presence of increased number of coagulated rubber clots in the film deposited after 15 seconds additional stirring.

Result: The mechanical stability value for the latex sample is expressed as the number of seconds elapsed from the start of the test to the end-point. Duplicate tests should check within 5% error.

D. Determination of Total Alkalinity

Apparatus: Electronic balance, burette, weighing bottle, glass electrode, pH meter.

Materials: Rubber latex, 0.1N HCl, methyl red indicator.

Sample Preparation: Total alkalinity of the latex is determined as per ASTM D 1076 method. Place approximately 5 g of latex into a glass-weighing bottle of 10 ml capacity with a ground cap, and weigh. Pour the sample into a beaker containing approximately 300 ml of distilled water, restore the cap of the bottle quickly to prevent the loss of ammonia, and set aside for weighing. The sample mass is equal to the difference between the two weighing. The transfer of the sample in to the beaker shall be done in such a way that no latex runs down the outside of the weighing bottle.

Procedure: Insert the electrode of a calibrated glass electrode pH meter into the liquid and note the pH. Slowly, and with continuous stirring add 0.1N HCl until a pH of 6.0 is reached. As the pH of 6.0 is approached, smaller increments should be added. Calculate the total alkalinity, reported as NH_3, based on grams of ammonia per 100 g of latex as follows:

$$\text{Total alkalinity as } NH_3 = \frac{1.7 \times N \times V \times 10}{W(100 - TSC)}\%$$

where, N is normality of standard HCl, V is volume of HCl added in ml and W is original mass of the latex, in grams.

If total alkalinity is required based on the water phase of latex, calculate as follows:

$$\text{Total alkalinity, as KOH} = \frac{56.1 \times N \times V \times 10}{W(100 - TSC)}\%$$

where, N is normality of HCl, V is volume of HCl added, W is weight of the latex, TSC is total solid content, 56.1 is the equivalent weight of KOH.

Result: Total alkalinity of the latex = _____ %

E. Determination of KOH Number

The KOH number of rubber latex is characterized as per ASTM D 1076-88 method.

Apparatus: pH meter, hot plate cum magnetic stirrer, electronic balance, 500 ml beaker.

Materials: Rubber latex, 5 % formaldehyde solution, 0.5 N potassium hydroxide solution.

Procedure: Standardize the pH meter with a standard buffer of known pH (pH – 9.2). Weigh out accurately a sample of latex containing approximately 50 g of solids in a 500 ml beaker. Adjust the ammonia content to 0.5% on water basis by addition of 5% formaldehyde solution (1 ml of 5% formaldehyde = 0.0189 g of NH_3) while stirring. Calculation for the amount of formaldehyde solution for this adjustment is as follows.

Formaldehyde solution (5%), ml

$$= \frac{W \times (100 - TS) \times (\%\ NH_3 \text{ in water phase} - 0.50)}{189}$$

where, *W* is the gram of wet latex sample, and *TS* percentage of total solids.

Weight of latex taken (*W*) = 50 × (100/*TS*)

$$\text{Percentage of } NH_3 \text{ in water phase} = \frac{\%\ \text{of } NH_3 \times 100}{100 - TSC}$$

Add enough distilled water to dilute the latex to about 30 % TS and insert the electrode and slowly add 5 ml of 0.5N KOH while stirring. After 10 s record the pH. Continue addition of 1ml increments of 0.5N KOH solution while stirring and record the pH after waiting 10 s after each addition. The end-point of the titration is the point of inflexion of the curve of pH value versus volume in ml of KOH solution. At this point, the first differential, reaches a maximum and the second differential changes from a positive to a negative. To determine the first differential, tabulate the difference in pH for each increment of 1 ml. Determine the second differential by the differences of the first differential again using the mean of the two values used for the difference. The point of inflexion is where the linear intercept crosses the zero line of the second differential.

Vol of 0.5N KOH in ml	*pH*	*Second differential*
5	10.12	0.01
6	10.13	0.01
7	10.32	0.02
8	10.44	0.02
9	10.58	0.05
10	10.77	0.06
11	11.04	0.05
12	11.30	–0.06
13	11.56	–0.06
14	11.74	–0.04
15	11.88	–0.04

Actual titre value = 11.0 + [0.05/(0.05 + 0.06)] = 11.45 ml

Calculate the KOH number, which is expressed as the number of grams of KOH required to neutralize the acids in 100 g of solids in latex, as follows:

$$\text{KOH Number} = \frac{V \times N \times 56.1}{TS \times \text{Weight of sample}}$$

where, V is volume of 0.5 M KOH, N is normality of KOH solution, and TS is percentage of total solids.

Result: Report the KOH number of the latex.

F. Determination of Viscosity of Latex by Brookfield Viscometer

Viscosity of rubber latex is measured as per ASTM D 1076 specifications.

Apparatus: Water bath, thermometer, 500 ml beaker, calibrated Brookfield viscometer, 180 micrometer sieve (Fig. 3.14).

Materials: Latex

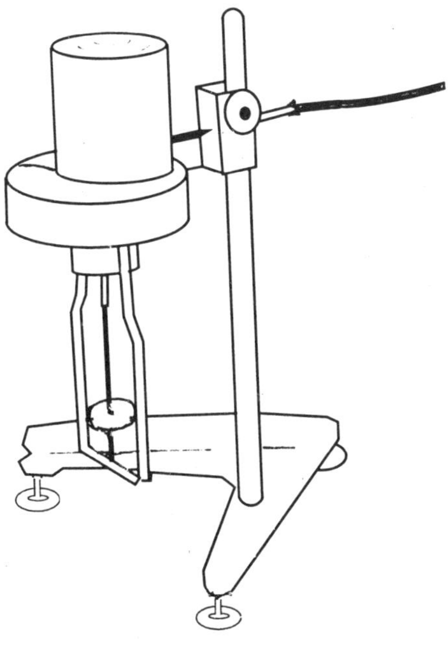

Fig. 3.14: *Brookefield viscometer*

Sample Preparation: Strain the sample latex through 180 micrometer sieve and adjust the latex to 60 $\pm$ 0.1 % TSC and condition to 25°C for 24 hr.

Procedure: Pour the prepared latex slowly down the side of a 500 ml beaker preventing incorporation of air. Attach the spindle No. 2 to the viscometer and then insert it into the latex until the surface of the latex is within the notch in the shaft of the spindle. Level the viscometer so the spindle is plumb.

Set the speed control to 60 rpm in Brookfield viscometer. Switch on the instruments and allow the spindle rotation to reach the equilibrium. Take the reading on the dial reading. Repeat the above procedure for different spindles and with different spindle speeds. Report the average values of the same.

Calculation: Multiply the dial reading on the scale with the spindle factor.

Result: The viscosity value is reported along with spindle number used in milliPascal second (mPa.s) which is equivalent to centipoise.

G. Determination of Coagulum Content

Outline: The material comprising pieces of coagulated rubber, latex skim and coarse foreign matter, retained on the sieve when filtered the latex through an 80 mesh stainless steel sieve.

Apparatus: Stainless steel sieve 80 mesh (180 micron), hot air oven, desiccator, analytical balance, 500 ml beaker.

Materials: Latex, potassium oleate solution (5%).

Procedure: Weigh out 200 ± 1 g of the well mixed latex sample into 500 ml beaker. Add 200 ml of potassium oleate solution and mix thoroughly. Weigh a clean and dry sieve (W_1). Wet the sieve with potassium oleate solution and then filter the latex-potassium oleate mixture through the sieve. Wash the residue on the sieve with potassium oleate solution until it is free from latex. Then wash the residue with clean water till the filtrate is neutral. Dry the sieve with coagulum for 30 minutes at 105 ± 2°C and cool in a desiccator and weigh (W_2). Calculate the coagulum content of latex using the below relation:

$$\text{Coagulum Content} = \frac{(W_2) - (W_1)}{2} \%$$

where, W_1 = Weight of empty sieve, W_2 = Weight of sieve with residue.

Results: Report the Coagulum content of the latex in %.

H. Determination of Magnesium in Latex

Outline: A known weight of the given sample is diluted and titrated against EDTA in presence of a buffer and an indicator. From the titre values the magnesium content in latex is calculated.

Apparatus: Conical flask, pipette, rubber latex, electronic balance, burette, 500 ml beaker.

Materials: Rubber latex, 0.05 M EDTA (1.86 g/l), 0.05 M $MgSO_4$ (1.2324 g/l), NH_4Cl/NH_4OH buffer (dissolve 67.5 g NH_4Cl in 200 ml water, add 570 ml NH_4OH and dilute to 1 litre), Eriochrome Black-T indicator (grind together 0.2 g Eriochrome Black-T and 50 g AR KCl).

Procedure: Take about 10 g of the well-mixed latex sample into a clean and dry weighing bottle and weigh (W_1). Transfer about 5 g of the sample into a 100 ml standard flask and reweigh the bottle (W_2). Dilute the latex with 100 ml water. Take around 300 ml water in a 500 ml conical flask, add 10 ml buffer solution and a pinch of Eriochrome Black-T indicator. Pipette out 20 ml of the diluted latex into the conical flask, mix well and immediately titrate against the 0.05 M EDTA solution. End-point is change in colour from wine red to light blue. Carryout the experiments in duplicate or until get the concordant values. Calculate the magnesium content using the following equation:

Wt. of magnesium in latex (mg/l)

$$= \frac{24.32 \times V \times M \times 1000 \times 100}{(W_2 - W_1) \times 20}$$

where, M is molarity of EDTA, V is volume of EDTA, $(W_2 - W_1)$ is weight of the latex taken.

Result: Report the magnesium content in the latex in mg/l.

Table 3.8: *Some typical properties of common solvents*

Solvents	*Chemical formula*	*Mol. volume (cm^3/mol)*	*B.P. (°C)*	*Visco-sity (mPa. .s)*	*Density (g/cc)*	*Solu-bility para-meter $(cal/cm^3)^{1/2}$*	*Dipole moment (debye)*
n-Pentane	C_5H_{12}	115.2	36	0.23	0.625	7.10	0.02
n-Hexane	C_6H_{14}	131.6	69	0.29	0.660	7.27	0.02
n-Heptane	C_7H_{16}	147.5	98	0.40	0.683	7.43	–
n-Octane	C_8H_{18}	163.5	126	0.52	0.704	7.57	0.02
n-Nonane	C_9H_{20}	179.7	151	0.67	–	7.69	–
n-Decane	$C_{10}H_{22}$	195.9	174	0.86	–	7.72	–
Methylene chloride	CH_2Cl_2	64.5	39.5	0.4	1.324	9.70	1.36
Chloroform	$CHCl_3$	80.7	61.18	0.54	1.487	9.20	1.02
Carbon tetrachloride	CCl_4	97.1	77.0	0.9	1.592	7.28	0.00
1,2-Dichloro-ethane	$C_2H_4Cl_2$	79.4	83	–	1.251	9.90	1.10
Tetrachloro-ethylene	C_2Cl_4	102.7	221	1.46	1.50	9.30	0.00
Benzene	C_6H_6	89.4	80	0.0647	0.879	9.2	–
Toluene	$C_6H_5CH_3$	106.9	111	0.059	0.867	8.9	–
Methanol	CH_3OH	40.4	65	0.0593	0.792	14.5	–
Ethanol	C_2H_5OH	58.2	78	0.012	0.791	12.7	–
Propanol	C_3H_7OH	74.6	97	0.1720	0.804	11.9	–
Water	H_2O	18.0	100	0.8007	1.000	23.4	–
p-xylene	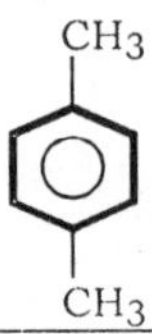	123.9	138	–	0.865	8.77	0.02

contd. ***Table 3.8***

Solvents	*Chemical formula*	*Mol. volume (cm³/mol)*	*B.P. (°C)*	*Viscosity (mPa. .s)*	*Density (g/cc)*	*Solubility parameter (cal/cm³)½*	*Dipole moment (debye)*
Chlorobenzene	C_6H_5Cl	102.2	132	0.0803	1.106	9.68	1.62
Methylethyl ketone	$CH_3—CO—C_2H_5$	90.2	80	–	0.806	9.30	2.76
Acetone	CH_3COCH_3	74	56	0.0295	0.791	10.0	2.69
Cyclohexanone	$C_6H_{10}O$	104.7	155	–	0.905	9.90	3.8
Acetonitrile	$CH_3—CN$	52.9	82	–	0.776	24.77	–
Acrylonitrile	$CH_2=CH(CN)$	66.3	77	–	0.804	21.50	–

Table 3.9: *Concentrations of commercially available acids*

Regent	*% by mass*	*Spec. gravity*	*Normality (N)*	*Molarity (M)*	*Volume req. to make 1 lt. of 1M solution*	*Volume req. to make 1 lt. of 1 N solution*
Hydrochloric acid (HCl)	35	1.18	11.3	11.3	89	89
Nitric acid (HNO_3)	70	1.42	16.0	16	63	63
Sulphuric acid (H_2SO_4)	96	1.84	36.0	18	56	28
Phosphoric acid (H_3PO_4)	85	1.69	41.1	13.2	69	23
Acetic acid (CH_3COOH)	99.5	1.05	17.4	17.4	58	58
Aq. ammonia (NH_4OH)	72	0.90	14.3	14.3	70	70

Table 3.10: *Details of dilute solution (2N) preparation*

Solution to be prepared	*Amount of acid or base to be added in water*
Dilute sulfuric acid	5 ml concentrated acid (d = 1.84 g/cc) in 90 ml water
Dilute nitric acid	13 ml concentrated acid (d = 1.40 g/cc) in 80 ml water
Dilute hydrochloric acid	19 ml concentrated acid (d = 1.16 g/cc) in 80 ml water
Dilute acetic acid	12 ml glacial acetic acid in 88 ml water
Dilute ammonia	17 ml concentrated ammonia (d = 0.882 g/cc) in 90 ml water
Dilute sodium hydroxide	8 g sodium hydroxide in 100 ml water

Table 3.11: *Molecular formula, formula weight and equivalent weight of some of the chemicals/salts*

Name of the compound	*Molecular formula*	*Formula (Molecular) weight*	*Equivalent weight*
Ammonium ferrous sulphate	$(NH_4)_2SO_4 \cdot FeSO_4 \cdot 6H_2O$	392.14	Formula weight
Copper sulphate	$CuSO_4 \cdot 5H_2O$	249.68	Formula weight
Ferrous sulphate	$Fe_2SO_4 \cdot 7H_2O$	278.02	Formula weight
Ferric chloride (anhydrous)	$FeCl_3$	162.21	Formula weight
Iodine	I_2	253.81	1/2 Formula weight
Oxalic acid	$H_2C_2O_4 \cdot 2H_2O$	126.7	1/2 Formula weight

Name of the compound	*Molecular formula*	*Formula (Molecular) weight*	*Equivalent weight*
Potassium hydrogen phthalate	$C_8H_4O_4K$	204.23	Formula weight
Potassium dichromate	$K_2Cr_2O_7$	294.19	1/6 Formula weight
Potassium permanganate	$KMnO_4$	158.04	1/5 Formula weight
Sodium chloride	NaCl	58.44	Formula weight
Sodium carbonate	Na_2CO_3	105.99	1/2 Formula weight
Sodium thiosulphate	$Na_2S_2O_3 \cdot 5H_2O$	248.18	Formula weight
Sodium hydroxide	NaOH	40	Formula weight

EXERCISE

1. Define the following terms and mention their significance: Density, bulk density, bulk factor, true density, apparent density, acid value, iodine value, aniline point, saponification value, viscosity, dynamic viscosity, kinematic viscosity, fire point, flash point, intrinsic viscosity, reduced viscosity, relative viscosity, etc.
2. What is epoxy? Explain the term epoxy equivalent weight (EEW). Mention its significance.
3. Mention the units of EEW. How many types of epoxies available commercially? Mention the uses of epoxy resin.
4. What is hydrolyzable chlorine content? What inference we can get from hydrolyzable chlorine content of an epoxy?
5. List out the different types of carbon black available commercially.
6. How do you characterize carbon black?
7. Explain the term DBP value. Mention its significance. How do you nomenclature the carbon black on the basis of DBP value?

8. Mention the factors affecting on surface area of carbon black.
9. Mention the different kinds of filler.
10. Explain the applications of carbon black.
11. How do you classify the oil on the basis of aniline point?
12. Name the different grades of processing oil available commercially.
13. Mention the uses of oil in polymer industry.
14. What is plasticizer? Mention its role in polymer field.
15. Explain the classification of plasticizer.
16. What are phthalates? Where they are extensively used?
17. Which additives extensively used in food packaging industry?
18. Mention the different compound ingredients used in rubber compounds? Explain their role.
19. What are the ingredients are used for PVC? Discuss their role and their drawback.
20. In how many ways molecular weight of polymer can be expressed? Name the methods of measurement of each one.
21. Define the term polydispersity. Why polydispersity of synthetic polymer is greater than one?
22. Explain the term MWD curve. Explain its significance.
23. What is Mark-Hownik equation? Which average molecular weight can be obtained by viscometric method?
24. State the basic principle of viscometric method to determine the $\overline{M}_v$.
25. Which average molecular weight can be obtained by end group analysis? Mention its limitations.
26. What is the industrial importance of cellulose acetate (CA)? On the basis of acetyl value how do you classify the cellulose acetate?
27. Explain the role of acetyl value on the properties of CA.
28. How do you convert the cellulose to cellulose acetate? Explain.
29. Mention other chemically modified products of cellulose.
30. List the uses of formaldehyde in polymer field. Why do you estimate formaldehyde content?
31. Mention the general uses of formaldehyde.
32. Does the determination of formaldehyde is a redox titration?
33. Name the important polymer prepared using formaldehyde.
34. List out the general uses of hydrogen peroxide.

35. Define the term saponification value. Mention its significance.
36. Mention the advantages and disadvantages of Saybolt viscometer method.
37. What is the significance of flash and fire point values?
38. Resin/paraffin wax/polymer is it having flash and fire point?
39. How do you measure the concentration of organic additives present in plastic/rubber (solvent extraction method)?
40. State the principle involved in solvent extraction method. Mention the advantages and disadvantages of solvent extraction method.
41. Distinguish between (i) PVC and PPVC; (ii) PS and HIPS.
42. Explain the classification of reinforcement.
43. Define the term polymer composite.
44. How do you measure the fibre content? Explain.
45. Explain the factors affecting on the fibre reinforced composites.
46. Give some examples for nitrogen containing and halogen containing polymers.
47. How do you find the detailed molecular weight distribution of the polymer without fractionating it?
48. Explain why colligative property measurements give number average molecular weight.
49. What structural parameters influence the melting point of the polymers?
50. Define the term threshold molecular weight.
51. Explain why in a polyhydrocarbon, such as polyethylene, threshold molecular weight is much higher than that for a polyamide.
52. How does data on the density of plastics help plastic industries?
53. Which combinations of miscible liquids are suitable for preparing density graident columns for polyolefins?
54. How is density gradient column prepared?
55. What are the advantages of using a gradient column for density measurement?
56. If the powder particles of a polymer are separated by sieving into small and large particle of uniform sizes, in what way does their apparent density differ from that of the original powder mix ?
57. What is the basis of designing standard fire tests?

58. Describe the course of a fire in plastic materials.
59. Discuss the significance of fire testing.
60. Compare the different types of standard test methods used for determining rate of burning.
61. Describe the process of smoke formation and the hazards related to it.

4 Identification of Polymers

INTRODUCTION

It is very important to identify and characterize polymers for better quality control and to maintain standards. Polymer samples to be identified will usually be in the form of pellets, flakes, powder, film, rod, tube or finished articles. It is very difficult to completely identify some plastics in the compounded or blended form. Hence, the complete identification involves physical tests, chemical detection, spectroscopic examination, extraction, etc. and is a vast subject by itself. Here, an attempt has been made to identify polymers by simple chemical methods.

A systematic chemical analysis for identification of polymeric materials is perhaps the only methods of true identification. However, some simple tests are also given for quick tentative checks. The best results are obtained when each of the step is followed correctly and when the polymer to be identified is simple.

Based on these a brief note on polymers has been described below.

A. RUBBERS

1. Natural Rubber: Mainly consists of cis 1,4-polyisoprene with minor proportions of other isomeric structures. Both the cis- and trans- isomers occur in nature as natural rubber (NR) and gutta-percha, respectively. Natural rubber is highly soft and elastic. It is soluble in carbon disulphide and petrol. Gutta-percha on the other hand, is a hard thermoplastic and dissolves in petrol only on heating and precipitates on cooling.

2. Chlorosulphonated Polyethylene: This rubber is a modification of polyethylene where crystalline thermoplastic is converted to an amorphous elastomer by chlorosulphonation. Polyethylene with substituted chloride and sulphonyl chloride groups, commonly contains about 27% chlorine and 15% sulphur. It is a polar rubber which imparts excellent chemical resistance and moderate oil resistance. As there is no unsaturation, it imparts excellent weather and age resistance. Moreover, the structure also shows that it cannot be vulcanized by a conventional sulphur and accelerator curing systems. Sulphonyl chloride groups provided reactive centres for vulcanization.

3. Ethylene-Propylene Rubber: This is a copolymer of ethylene and propylene, commonly with about 30% propylene. This rubber generally contains a small amount of diene.

4. Ethylene-Propylene Diene Terepolymer (EPDM): It is the terepolymer which is polymerized from ethylene, propylene and a small percentage of diene which provides unsaturation in side-chain pendant from the saturated backbone. It has a greater oil and solvent resistance, better heat, oxygen and ozone resistance. It has a narrow molecular weight distribution, a fair tear resistance, very good abrasion and impact resistance. Their environmental resistance is also outstanding. They are used in electrical insulated jackets, footwears, fire proofed fabrics and automotive moulded parts.

5. Butyl Rubber: Butyl rubber is marketed as a copolymer (IIR rubber) containing about 0.5% to 2% of isoprene, which provides the necessary unsaturation sites for vulcanization. Butyl rubber exhibits outstanding resistance to attack by oxygen, ozone and chemicals. It exhibits low gas permeability and high hysteresis loss. For these reasons, it finds principle usage in weather strippings, adhesives, chemical condenser packings, high temperature conveyor belts, mounts for appliance motors, vibration absorbers for bridge, etc. High-molecular-weight butyl rubber may be prepared by low-temperature cationic polymerization using Lewis acids like aluminium chloride in chlorocarbon solvents such as chloromethane.

6. Halobutyl Rubbers

(*i*) *Fluorinated Rubbers:* Various fluorine-containing rubbers are available. Most commonly they are copolymers of vinylidene fluoride with chloro trifluoro- ethylene or perfluoro propylene.

(*ii*) *Chlorobutyl rubber (CIIR):* Chlorobutyl rubber is produced by introducing small amount of chlorine into the butyl rubber. It is used in inner liners of tubeless tyres, tyre curing bags, pharmaceutical stoppers and conveyor belts. It is characterized by low permeability to air, heat resistance and flux resistance.

(*iii*) *Bromobutyl rubbers (BIIR):* It is similar to chlorobutyl rubbers but it has faster cure rate. Bromobutyl is used in inner linings, side-wall, barrier strip compounds for tyres and in heat resistant inner tubes.

7. Acrylate Rubber: Acrylic rubbers are co-polymers of acrylic esters which are monomers, containing small amounts of chlorine. These are high heat and oil resistant rubbers. They belong to the class of speciality elastomers. Acrylics exhibit good damping qualities and can be compounded to provide excellent flex life, fair electrical insulation resistance and some degree of flame. Acrylic rubbers are widely used in automobile industries.

8. Acrylonitrile-Butadiene Rubber (NBR): It is commonly known as *Nitrile Rubber.* Copolymers of butadiene and acrylonitrile, commonly with 20–40% acrylonitrile. It has excellent properties like; resistance to aliphatic hydrocarbon oils, fuels and greases; resistance to heat; low gas permeability and low permanent set. For most of its applications NBR is employed because of its resistance to hydrocarbon fuels, oils and greases.

9. Polybutadiene Rubber: It is mainly cis- and trans- 1,4-polybutadiene with minor 1,2- contents. Cis-1,4-polybutadiene is a rubber and 1,2-polybutadiene is a thermoplastic. Cis-polybutadiene have very high abrasion resistance and hence, find applications in automobile tyres. PB as such, however, are not popular synthetic rubbers, but co-polymers of butadiene and styrene (SBR) or those of butadiene and acrylonitrile (nitrile rubber) are two commercially important synthetic rubbers.

10. Polychloroprene Rubber: It may be a homopolymer or copolymer with small amounts of co-monomer. It may or may not be sulphur modified. Polychloroprene is produced from chloroprene. This elastomer is principally used for providing oil-resistant insulation coatings to wires and cables and manufacturing of shoe soles, gloves and industrial hoses.

11. Polyisoprene Rubber: It is a synthetic material closely resembling natural rubbers. It has excellent physical properties including abrasion and low temperature resistance. It has poor resistance to petroleum based fluids.

12. Polyurethane (PU) Rubber: These are polymers formed by reaction of di-isocyanates and polyesters, polyether or polyester amides. The abrasion resistance of PUs has been recognized as one of their outstanding attributes. It has good low temperature flexibility. PU shows superior resistance to degradation by oxygen and by ozone and to swelling by hydrocarbon fuel and oils. Finished PU products appear in the solid, microcellular or cellular, thermoset and thermoplastic, flexible, elastic and rigid status. Applications include moulded and extruded goods, binders, adhesives, films, coatings, jackets, prostheses, exterior automotive parts, furniture, sports equipment and so forth.

13. Silicone Rubber: Silicone polymers are thermally stable. Silicone elastomers have high-temperature and oxidative stability, low temperature flexibility, good electrical properties, and resistance to weathering and oil. They are available in liquid, waxy and rubbery forms. Silicone rubbers are used as elastomers and can be vulcanized with the help of peroxide initiator. The important applications include wire and cable insulation, surgical implants, gasket materials and seals and aircraft tubing.

14. Styrene-Butadiene Rubber (SBR): Copolymers of styrene and butadiene, commonly with about 30% styrene (high styrene resins containing about 85% styrene are also available). SBR, which is commercially known as GR-S or Buna-S, was the most important rubber in the 1970s as it occupied about half of the global consumption of rubber compounds. This was because of its low cost, good abrasive resistance and higher level of product uniformity than natural rubber.

15. Vinyl Pyridine Rubber: Various vinyl pyridine-containing rubbers are available. Most common one is a terepolymer with butadiene and styrene.

16. Polysulphide Elastomer: Alkyl halides can react with alkali polysulphides to give polysulphide elastomers. Polysulphides have excellent oil resistant properties and are linear condensation polymers. The polysulphides are usually cured with the help of lead dioxide and, unlike rubbers, sulphur does not vulcanize them. Sulphur, however, can increase the rate of the curing reaction. Carbon black is frequently employed as a reinforcing material for polysulphide. Polysulphides are mainly used to make sealants, gaskets, balloons, fabric coatings and gasoline hoses. They have excellent oil-resistance and very low gas permeability and are practically unaffected by ozone. Also, aging is very slow in case of this polymer. Polysulphides form excellent solid propellants for rockets upon mixing with inorganic oxidizers such as ammonium perchlorate.

B. FLEXIBLE THERMOPLASTICS

1. Polyethylene (PE): PE is a homopolymer with various degrees of chain branching and hence degrees of flexibility. Different kinds of polyethylenes commercially available are: LDPE, HDPE, LLDPE, MDPE, HMHDPE, UHMWPE, XLPE, etc. LDPE melts at 110–125°C and is only around 40% crystalline. The density is around 0.91–0.92 g/cc. While practically, no solvent dissolves it at room temperature, several solvents can do so at high temperatures. HDPE is 90% crystalline and its density is around 0.94–0.96 g/cc. It has a melting point in the range of 135–150°C.

They find use in the manufacture of toys, pipes, packaging films, household articles, etc.

2. Polyisobutylene: This has short-term rubbery properties but cannot be vulcanized. Polyisobutylene exhibits low gas permeability, high hysteresis (high damping), and superior resistance to heat/ozone/oxidation, weather and chemical substances.

3. Polyvinyl Alcohol (PVA): It is a water-soluble polymer and extensively used as wet strength adhesives. PVA is manufactured by the hydrolysis of PVAc in the presence of acids or alkalis. Various amounts of residual acetate groups may be present. PVA fibres have excellent dimensional stability.

4. Polyvinyl Chloride (PVC): PVC is one of the cheapest and most widely used plastics globally. The straight polymer is rigid, but is very commonly plasticized to give a flexible material. May also be a copolymer with small amounts of various comonomers. It is used for large scale production of cable insulation, equipment parts, pipes, laminated materials and in fibre manufacturing.

C. RIGID THERMOPLASTICS

1. Acrylonitrile-Butadiene-Styrene (ABS): ABS is a copolymer of three monomers such as —acrylonitrile, butadiene and styrene. Copolymers of various composition and structure can be obtained depending on the method of manufacturing.

2. Cellulose Acetates (CA): Different kinds of cellulose acetates available commercially are: cellulose acetate, cellulose secondary acetate (CSA) and cellulose triacetate (CTA). They are used to make moulded decorative articles, fountain pens, frames for spectacles, etc.

3. Cellulose Acetate-Butyrate (CAB): Esters with various proportions of acetate and butyrate groups, commonly 20–40% butyrate.

4. Cellulose Nitrate: Cellulose nitrate also known as nitrocellulose (NC), is a nitric acid ester of cellulose. In plastic materials about 60% of the available hydroxyl groups are nitrated. Depending on the degree of nitration, cellulose nitrate has different applications.

5. Ethyl and Methyl Cellulose: Ethyl cellulose is flexible and tough even at very low temperatures. It has better mechanical and electrical properties compared to other cellulosic plastics. Both methyl and ethyl cellulose find the applications as finishing materials (in the textile industries) and as paint bases.

6. Nylons: The aliphatic polyamides are generally known as *nylons*. There are different types of nylons usually indicated by a numbering system based on the number of the carbon atoms present in the monomer molecule. Most commonly or commercially available types are: nylon-6; nylon-6,6; nylon-6,10; nylon-4,6; nylon-11 and nylon-12.

7. Polyacrylonitrile (PAN): Addition polymerization of acrylonitrile yields PAN. Widely PAN is commercially available in the form of copoymers rather than homopolymer. PAN is most often encountered as a fibre.

8. Polycarbonate (PC): Polyesters derived from carbonic acid derivatives and dihydroxy compounds, commonly from phosgene and 4,4- dihydroxy diphenyl -2,2-propane. Almost all PCs used as plastics contain bisphenol A. It is a transparent engineering plastic.

9. Polychloro Trifluoro Ethylene (PCTFE): Flame retardant fluorinated polymer.

10. Polyethyleneterephthalate (PET): They are prepared by a polycondensation reaction between a dicarboxylic acid (terephthalic acid) and a diol (ethylene glycol). PET melts at around 265°C and is resistant to heat and moisture. This polyester is virtually unattacked by many chemicals. It is extensively used to make textile fibres. It has good mechanical strength upto 175°C. Garments made from its fibres resist the formation of wrinkles.

11. Polybutylene terephthalate (PBT): It is a condensation polymer of terephthalic acid and butylene glycol. It is a transluscent polymer and widely used in packaging industries.

12. Polyformaldehyde or Polyacetal or Polyoxymethylene (POM): It may be a homopolymer or copolymer with small amounts of comonomer.

13. Polymethyl Methacrylate (PMMA): It is a colourless transparent plastic with excellent outdoor life period and good strength. It is amorphous by nature, owing to the presence of bulky side group in the molecule. Optical clarity is the main feature of this plastic. In many applications, it is an excellent substitute for glass. It also has good mechanical properties. However, compared to glass, it has poor scratch resistance. PMMA is used to make attractive sign boards and durable lenses for automobile lighting. It is also used in building for decorative purpose.

14. Polypropylene (PP): PP is perhaps the lightest known industrial polymer. Polypropylene may have various degrees of tacticity. It is transparent and a low density (0.88–0.90 g/cc) polyolefin polymer. Three kinds of PP are available a commercially. They are isotactic, syndiotactic, and atactic PP. It is used in appliances like refrigerators, radios, TVs, etc. It is also used in packaging films, pipes, seat covers, ropes, car bumpers, etc.

15. Polystyrene (PS): PS is also known as polyvinyl benzene. PS can be prepared by addition polymerization. It is transparent and brittle polymer. This is available commercially in various forms. PS is chemically inert. PS is very widely used in the manufacture of articles such as moulded containers, lids, jars, bottles, radio and television cabinets, toys, foamed plastics and many other household items.

16. Polytetrafluoro Ethylene (PTFE): PTFE is also known as Teflon (trade name). It is used in non-lubricated bearings and its fibre is used to form belts, filter cloth, where resistance to acids and alkalis are required. It is a heat resistant fluoro-polymer.

17. Polyurethane (PU): Polymers prepared from di-isocyanates and macrodiol alcohols, commonly 1,6-hexamethylene di-isocyanate (HDI) and 1,4-butanediol. PUs are used to make foams, coatings, adhesives and elastomers.

18. Polyvinyl Acetate (PVAc): Addition polymers of vinyl acetate yields PVAc. Along with the other polymers and compounds, it is used in adhesives, lacquers and for the large-scale manufacture of gramophone records.

19. Polyvinyl Fluoride (PVF): PVF is a polar and symmetric fluoropolymer. PVF is crystalline polymer with a melting point of around 200 °C. It has excellent out door life. It is highly resistant to chemicals and in that respect resembles other fluoro carbon polymers such as Teflon. It maintains good structural integrity even above 150°C and has good thermal stability. Gas permeability is sufficiently low. It has already proved its merit as a protective coating as well as a glazing material over cement mortars in the building industry. For durability, even in extremely small thickness, it is better than the conventionally used materials such as paints and other surface coatings.

20. Polyvinylidene Chloride (PVDC): It may be a homo-polymer or more commonly, a copolymer with about 20% vinyl chloride. The PVDC polymer is prepared by radical polymerization using the emulsion technique. This polymer is insoluble in many solvents. Its main use is in making fibres. The vinylidene chloride homopolymer and its copolymers with vinyl chlorides, vinyl acetate or acrylonitrile, come under the special category of chlorine containing fibre forming polymers and are extensively used as fibres. Copolymers with ethyl acrylates, however, form thermoplastic materials.

21. Polyimides: Polycondensation of pyromellitic anhydride and p,p-diamino diphenyl ether results in the formation of polyimide. Polyimide finds extensive applications in electrical industries as the insulation coating of electromagnetic wirings. Polyimide can withstand temperatures upto 425°C for short-time exposures without undergoing any degradation or deformation. For this reason, the polymer finds applications such as surface coating in supersonic aircraft.

22. Polysulphone: Polysulphone is a condensation product of bisphenol A and dichloro diphenyl sulphone. It is a thermoplastic with a very high thermal stability and good mechanical strength over a wide range of temperatures (from −180 to +140°C). The extraordinary thermal stability arises out of resonance stabilization due to aromatic rings near —SO_2— groups. Polysulphone is used to make injection-moulded items and films.

23. Polylactic acid (PLA): PLA is one of the important biodegradable polyesters. PLA is polymerized through the self-polycondensation of lactic acid, which can be produced by the fermentation of potato waste. Since PLA breaks down in the environment back to lactic acid, which can be metabolized, it has found commercial use primarily in medical applications such as sutures, drug-delivery systems and wound clips. It may also have some agricultural applications, such as timed-release coating for fertilizers and pesticides and mulch films for moisture and heat retention and to reduce the weed population between rows of crops. Although PLA is relatively expensive compared with most commodity thermoplastics, recent research has shown that the monomer, lactic acid, can be produced from bacterially fermented starch obtained from food wastes such as potato peelings.

D. THERMOSETTING PLASTICS

1. Alkyds: These materials are of two distinct types, viz.; those based on unsaturated linear polyesters, which are subsequently cross-linked by reaction with an unsaturated compound and those which are based on diallyl phthalate (DAP). DAP compound's specific gravity ranges from 1.30 to 1.85, depending upon the type of filler. It burns with a yellow flame, produces black smoke and gives off the faint odour of phenol. It is self-extinguishing in nature and is difficult to ignite.

2. Epoxides: The epoxy polymers are basically polyethers. One type of epoxy polymer is prepared from epichlorohydrine and bisphenol A. The epoxy resins obtained through these reactions will be highly viscous liquids or solids with high melting points. The epoxy resins can be further cured with substances such as diamines, dicarboxylic acids, etc. It is to be noted that uncompounded epoxides being linear polymers, are thermoplastics.

The specific gravity of epoxy compounds range from 1.10 to 2.10. It burns with a yellow flame and gives off a black smoke that gives off a pungent amine odour. Infrared spectroscopy and thermal analysis can characterize different types of epoxies.

Epoxy resins find a large number of uses because of their remarkable chemical resistance and good adhesion. Epoxy resins are excellent structural adhesives. They are used in industrial floorings, foams, materials for electrical insulations, etc. One of the principal constituents in many of the fibre-reinforced plastics (FRP) is an epoxy polymer.

3. Melamine-Formaldehyde (MF): It is a condensation polymer of melamine and formaldehyde. It is a self-extinguishing thermoset have a specific gravity of 1.47–1.8. It burns with a yellow flame with a blue edge and gives off a fish-like odour. The material swells, cracks and turns white at the edges of a burned section. Melamine-formaldehyde is soluble in acetone and cyclohexanone.

4. Urea-Formaldehyde (UF): It is a condensation polymer of urea and formaldehyde. Urea-formaldehyde burns with a yellow flame with a greenish-blue edge. It gives off a strong odour of formaldehyde and the material usually swells and cracks as it continues to burn.

5. Phenol-Formaldehyde (PF): These are polymers based on phenols, commonly phenol or cresol, and formaldehyde. Hexamethylene triamine is commonly used in moulding powders and products may thus contain nitrogen. It is to be noted that uncompounded novolacs being linear polymers are thermoplastic. PF burns with a yellow flame and gives off the odour of phenol depending upon the type of filler. It may not be self-extinguishing. Phenolic compounds are only available in dark colours. The specific gravity lies in the range 1.30–1.90. Uncross-linked PF is soluble in acetone and acetic acid.

6. Polyesters: These are unsaturated linear polyester polymers cross-linked by various liquid vinyl monomers, commonly styrene. Polyesters burn with a blue edged yellow flame and produce black sooty smoke. They continue to burn without dripping after removal of the ignition source and give a cinnamon odour. The specific gravity ranges from 1.30 to 1.50.

7. Silicones: Silicones are produced in the form of resins, oils, greases, and also rubber-like elastic products. These materials also appear as processing aid in the manufacture of plastics, as impregnation compounds, coatings, separating membranes, mold releases, etc. Silicones burn with a bright yellow flame without odour and continue to burn after removal of the flame source. The specific gravity lies in the range from 1.05 to 2.82. The positive identification is made by an infrared spectrometer.

Classification of Engineering Materials

- METALS
  - **Basic**: Aluminum, Magnesium, Beryllium, Titanium, Copper, Nickel, Gold, Iron, Cadmium etc.
  - **Alloys**: Steel, Brass, Bronze, All metals which are not basic metals.
- NON-METALS
  - **Polymers**
    - **Thermosetting**: Alkyds, Dialyl Phthalates, Epoxies, MF, PF, RF, UF, CF, Polyurethanes, Urethanes (rigid foams), Silicones, (rigid resins), etc.
    - **Elastomers**: Acrylates, Butyls, Chlorosulphonated PE, Butadiene, SBR, EPDM, Fluorocarbons, Fluorosilicones, Polysulphides, PU, NBR, Neoprenes, Nitriles, Silicones
    - **Thermoplastics**: Acetates, Acrylics, Cellulosics, Chlorinated PE, Polycarbonates, Fluorocarbons, Nylons, Polysulphides, Polyethylenes, and PP, PPO, PPS, Polyimides, etc.
  - **Ceramics**: Aluminas, Beylias, Carbides, Cordierites, Forsterites, Lavas, Magnesias, Nitrides, Titanias, Steatites, etc.
  - **Glasses**: Silicas, Sodalimes, Lead, Borosilicates, Aluminosilicates, etc.
  - **Others**: Papers, Greases, Lubricants, Oils, etc.

Part A

Identification of Plastics

Polymers are broadly classified as thermoplastic, thermoset and elastomer. Five steps namely preliminary tests, elemental analysis, melting point, solubility test and chemical/confirmation tests are involved in systematic qualitative analysis of both plastics (thermoplastics and thermosets) and elastomers. They are briefly described in this chapter.

I. PRELIMINARY TESTS

1. Visual Appearance: Raw polymers are usually transparent or translucent or whitish or pale-brown. Thermoplastics are generally produced as granules or powders. Uncompounded thermoset resins are in the form of either highly viscous liquids or low temperature melting solids. Compounded plastics are generally coloured and opaque. Acrylics (PMMA, PMA, etc.), PS, PP, PC, cellulose acetate, SAN and some grades of PVC are clear transparent.

Phenolic resins are invariably dark in colour whereas, UF and MF resins are available in light colours. PTFE is milky white in colour.

2. Heating Test: Here, few granules or a bit of powder of the polymer is heated in a small spatula. The thermoplastics melt and the thermosets char giving pyrolysed products. To test finished articles, a hot soldering iron rod is kept over them. Thermoplastics soften and melt and thermosets degrade and char directly without melting.

3. Cutting Test: When cut with a sharp knife, thermoplastics can be easily pared off, whereas thermosets become flaky or powdery. PS and PMMA are exceptions in thermoplastics since they cannot be pared off.

4. Colour and Odour: Phenolic resins are invariably dark in colour, whereas UF and MF resins are available in light colours. PTFE is milky white or gray in colour. Virgin PET, acrylics, PC, PS, PAN and atactic PP are transparent. Nylon, PVC, HDPE, LDPE, LLDPE and PU are translucent in the uncompounded state. PET has a greenish tint, while SAN has a bluish tint. PPVC will have a smell of plasticizers, while cellulose nitrate gives camphor smell.

5. Floatation Test: If the plastic floats on the water, it may be polyolefin, usually PE or PP. But filled PE or PP sinks in water. PP and HDPE are hard to chew, while LDPE is soft to chew. PP is highly glossy in nature. Further, we can test the floatation on a solution of 70 g of hypo (sodium thiosulphate) in 60 ml of water. Those plastics with a specific gravity less than 1.2 sink here also. Flow chart 4.1 and Table 4.1 give the information on floatation behaviour and specific gravity of some plastics.

Flow Chart 4.1: *Floatation/Drop Test for Polymers*

Drop a Piece of Sample in Water

- SINKS → Drop on Hard Surface
  - Metallic Ring sound → Probably PC, PS, SAN, HIPS, ABS, PPS, PPO
  - Dull sound → Probably CA, CAB, PVC, Nylon, PTFE, PMMA, Acetals
- FLOATS → Either PE or PP → Scratch it with nail
  - a) Easily scratched; b) Low gloss → Probably LDPE
  - a) Can be scratched; b) High gloss → Probably HDPE
  - a) Cannot be scratched; b) High gloss → Probably PP

Table 4.1: *Polymer structure, specific gravity, glass transition temperature (T_g), melting temperature (T_m) and maximum useful temperature range of some plastics*

(A) Polymer Floats On Water

Plastic Materials	*Polymer Structure*	*Specific Gravity*	T_g (°C)	T_m (°C)	*Maximum Useful Temp. Range (°C)*
Polypropylene (PP)	$[—CH_2—C(CH_3)H—]_n$				
Isotactic PP (IPP)		0.90-0.91	−20	170	80–110
Syndiotactic (SPP)		0.90	−19	140	80–95
Atactic PP (APP)		0.89	−18	135	70–90
Polyethylene (PE)	$[—CH_2—CH_2—]_n$				
Very low density polyethylene (VLDPE)		0.91	−125	108	50–60
Low density polyethylene (LDPE)		0.91-0925	−130	105-115	50–60
Linear low density polyethylene (LLDPE)		0.92-0.93	−125	115-125	50–60
High density polyethylene (HDPE)		0.94-0.95	−130	120-135	60–80

High mol.wt. high density polyethylene (HMHDPE)		0.96	–132	135	70–80
Ultra high molecular wt. (UHMWPE) Polyethylene		0.97	–	140–145	70–80
Ethylene-vinyl acetate (EVA) copolymer	$(CH_2—CH_2)_x—(CH_2—CH(O—COCH_3))_y$	0.92-0.94	–20	95–103	50–77
Polymethyl Pentene (PMP)	$(CH_2—C(CH_3)(C_3H_7)—)_n$	0.83	15–55	245	121–160
(B) Sinks in Water and Floats on Hypo					
Polystyrene (PS)	$[—CH_2—CH(C_6H_5)—]_n$				
General Purpose PS (GPPS)		1.07	100	190	65–90
Medium Impact PS (MIPS)		1.05	95	185	65–90
High Impact PS (HIPS)	$[—CH_2—CH=CH—CH_2-]_X-[-CH_2—CH(C_6H_5)-]_Y$	1.02	93	185	65–90

(contd.)

Plastic Materials	*Polymer Structure*	*Specific Gravity*	T_g (°C)	T_m (°C)	*Maximum Useful Temp. Range (°C)*
Polychlorostyrene	$[—CH_2—CH(C_6H_4Cl—]_n$	<1.25	119	—	—
Poly-α-methyl styrene	$[—CH_2—C(CH_3)(C_6H_5)—]_n$	—	—	170/ 192	—
Poly Acrylonitrile (PAN)	$[—CH_2—CH(CN)—]_n$	1.17	170	265	70–80
Styrene-Acrylonitrile (SAN)	$[—CH_2—CH(C_6H_5)—]_x$ $—[—CH_2—CH(CN)—]_y$	1.10	115	200	80–100
Acrylonitrile Butadiene Styrene (ABS)	$—[—CH_2—CH(C_6H_5)—]_x—$ $—[—CH_2—CH{=}CH_2—CH_2—]_y$ $—[CH_2—CH(CN)]_Z—$	1.10	50	200	70–95
Polyvinyl Alcohol (PVA)	$—[CH_2—CH(OH)]_n—$	1.16-1.28	70-90	215-235	80-90
Polyphenylene Oxide (PPO)	$—[—O—C_6H_2(CH_3)_2—]_n—$	1.06	125	262	100–120
Polyvinyl Acetate (PVAc)	$—[CH_2—C(OCOCH_3)H—]_n$	1.17-1.20	80	175	50–70

Polyamides					
Nylon-6	[—NH—$(CH_2)_5$—CO—]$_n$	1.14	50	210–220	100–120
Nylon-6,6	[—NH—$(CH_2)_6$—NHCO—$(CH_2)_4$—CO—]	1.16	55	240–265	110–135
Nylon-11	[—NH—$(CH_2)_{10}$—CO—]$_n$	1.04	48	190–200	100–130
Nylon-12	[—NH—$(CH_2)_{11}$—CO—]$_n$	1.02	50	180–190	100–125
Nylon-6,10	[—NH—$(CH_2)_6$—NHCO—$(CH_2)_8$—CO—]	1.07	50	215–220	100–110
Nylon-4,6	[—NH—$(CH_2)_4$—NHCO—$(CH_2)_4$—CO—]	1.18	50	295	120–150
(C) Sinks In Both					
Polyacrylates					
Polymethylmethacrylate (PMMA)	—CH_2—C$(CH_3)_n$— with pendant $COOCH_3$	1.18–1.19	85–105	190	55–90
Polymethyl acrylate (PMA)	—{CH_2—CH$(COOCH_3)$}$_n$—	1.22	8	—	90
Polyethyl acrylate (PEA)	—{CH_2—CH$(COOC_2H_5)$}$_n$—	1.12	–22	—	40–80
Polypropyl acrylate (PPA)	—(CH_2—CH[$COOC_3H_7$])$_n$—	<1.08	–44	115/162	20–70

(contd.)

Plastic Materials	*Polymer Structure*	*Specific Gravity*	T_g *(°C)*	T_m *(°C)*	*Maximum Useful Temp. Range (°C)*
Polybutyl acrylate (PBA)	$—(CH_2—CH[COOC_4H_9])_n—$	1.0	–52	47	0–50
Polybutyl methacrylate (PBMA)	$—(CH_2—C(C_4H_9)[COOCH_3])_n—$	<1.0	—	—	10–60
Cellulose secondary acetate (CSA)	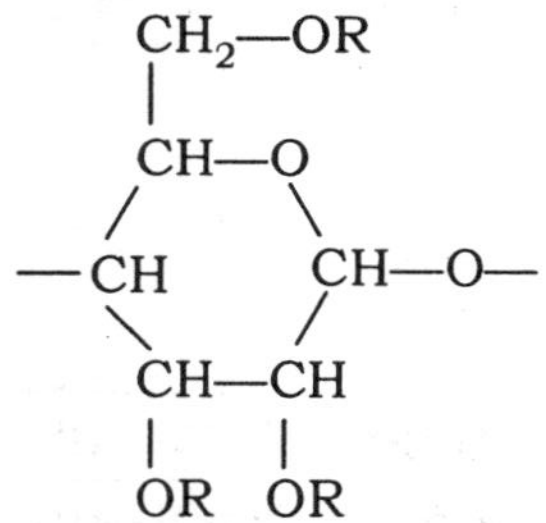 R = $COCH_3$	1.30	157	230	60–70
Cellulose acetate butyrate (CAB)	R = COC_3H_7	1.15-1.22	97	180-200	50–60
Cellulose nitrate (CN)	R = NO_2	1.60	127-135	200	55–65
Cellulose acetate propionate (CAP)	R = COC_2H_5	1.16-1.23	140	235	60–70

Polyvinyl chloride (PVC)	$[—CH_2—CH(Cl)—]_n$	1.40	70-80	155-170	65–80
Polyvinylidene chloride (PVDC)	$[—CH_2—CCl_2—]_n$	1.54-1.67	–4–15	160-190	80–95
Polyvinylidene fluoride (PVDF)	$[—CH_2—CF_2—]_n$	1.76-1.78	–20–40	172-185	140–150
Polyvinyl fluoride (PVF)	$[—CH_2—CHF—]_n$	1.38	40	200	100–110
Polychlorotrifluoro ethylene (PCTFE)	$[—CFCl—CF_2—]_n$	2.10	45-52	210	190–200
Polyurethane (PU) (Thermoplastic)	$[—O—(CH_2)_X—O—CO—NH—(CH_2)_Y—NH—]_n$	1.11-1.26	–51	160-180	80–90
Polytetrafluoro ethylene (PTFE)	$[—CF_2—CF_2—]_n$	2.13-2.30	–115	327	250–260
Polyoxymethylene (polyacetal)	$[—CH_2—O—]_n$	1.41-1.42	–13	165-175	80–90
Polycarbonate (PC)	$[—C_6H_4—C(CH_3)_2—C_6H_4—O—CO—O—]_n$	1.20-1.22	148	220	100–110
Polyethylene terephthalate (PET)	$[—OOC—C_6H_4—COO—(CH_2)_2—O—]_n$	1.37	75	255	130–140

(*contd.*)

Plastic Materials	Polymer Structure	Specific Gravity	T_g (°C)	T_m (°C)	Maximum Useful Temp. Range (°C)
Polybutylene terephthalate (PBT)	$[-OOC-C_6H_4-COO-(CH_2)_4-O-]_n$	1.32	30	225	120–130
Polyether imide (PEI)	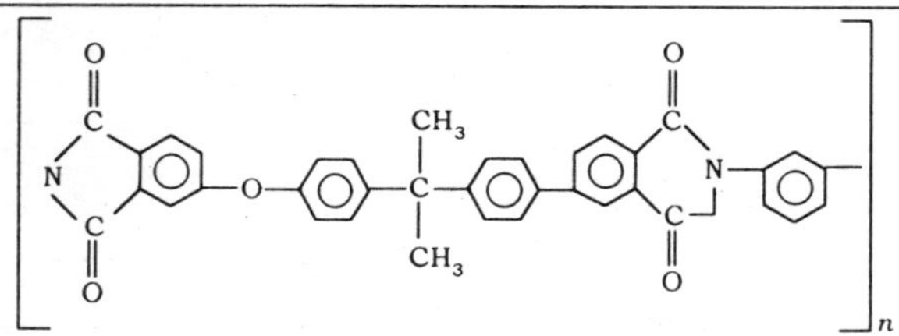	1.27	215–220	310	270–290
Polyamide imide (PAI)	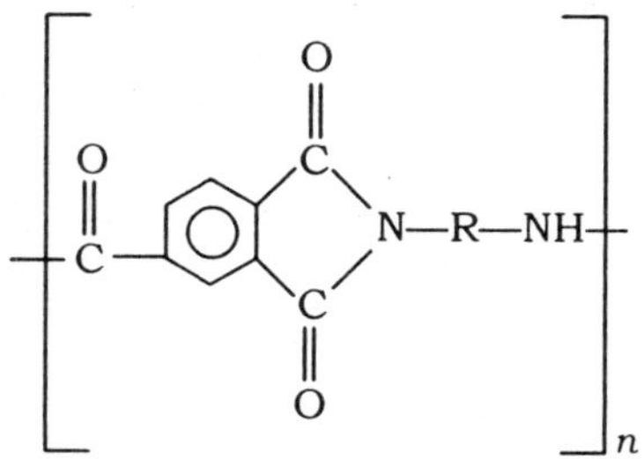	1.40	270–280	—	220

Polyimide (PI)	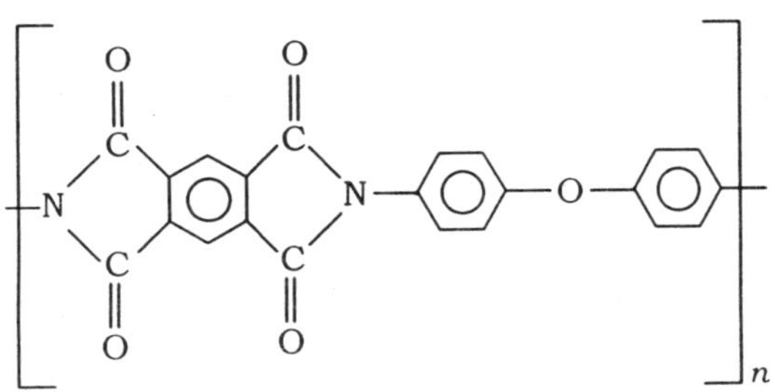	1.43	310–365	—	304
Polyacrylamide (PAA)	$-[CH-CH(CONH_2)]_n-$	—	160–170	240	220
Polysulphone (PSF)	$[-C_6H_4-SO_2-C_6H_4-O-C_6H_4-C(CH_3)_2-C_6H_4-O-]_n$	1.25	185-190	240	140–150
Polyphenylene sulfone (PPS)	$[-C_6H_4-S-]_n$	1.35	90	290	240–250
Polyether ether ketone (PEEK)	$-[C_6H_4-CO-C_6H_4-O-C_6H_4-O]-$	1.32	154	335-340	70–290

(contd.)

Plastic Materials	Polymer Structure	Specific Gravity	T_g (°C)	T_m (°C)	Maximum Useful Temp. Range (°C)
Fluorinated ethylene propylene (FEP)	$-[(CF_2-CF_2)-(CF_2-CF(CF_3))]_n-$	2.16	100	275	204
Polyaryl ether (PAE)		1.14	—	—	121
Polyether sulphone (PES)	$-[C_6H_4-SO_2-C_6H_4-O]-$	1.36	220–230	—	260
Polyaryl sulphone	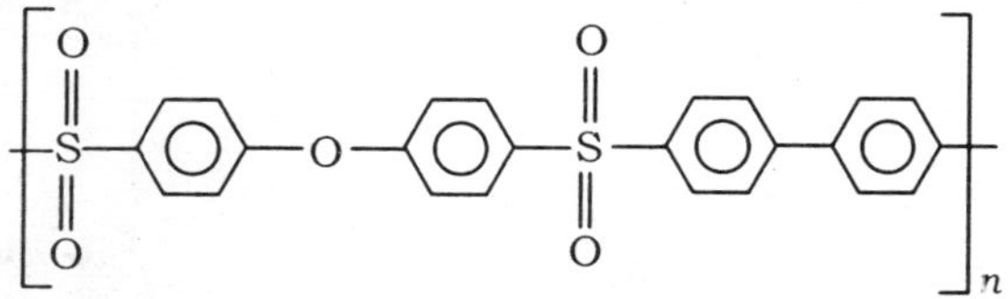	1.36	228	—	180
Silicone polymer (polydimethyl silicone (PDMS)	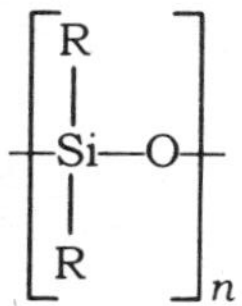	1.1-1.6	–123	–85 to –65	177
Poly butylene	$-(CH_2-CH(C_2H_5))_n-$	0.93-0.94	–20 to –25	120-124	80

6. Rigidity/bending Test: Some plastics exhibit particular characteristics in the manner in which they respond to bending. It is important that the specimens are of approximately the same dimension for getting better comparative results. Table 4.2 gives the bending behaviour of different plastics.

7. Film Tear Test: Many plastics are available as thin films and because of the method of production, they may exhibit specific characteristics in the manner in which they tear. Cut a piece of film to about 50 mm × 12 mm. Then make a short cut at one end with a- pair of scissors. Pull and attempt to tear the film rest of the way. Repeat the test in a direction at right angles. Record your observations and compare the results with Table 4.2.

Table 4.2: *Bending and Tearing Behaviour of some Plastics*

Plastic Materials	*Bending Behaviour*	*Tearing Behaviour*
Polyethylene (PE)	Bends, tends to remain	Tough, stretches before tearing
Polypropylene (PP)	Does not retains the bend	Tough, stretches before tearing
Polystyrene (PS)	Cracks, but retains the bend	Tears straight (bi-axially oriented)
PVC (Rigid)	Bends easily and springs back quickly	Stretches and then tears raggedly
Cellulose acetate	Bends, tends to remain	Tears easily and straight
ABS	Bends and tends to remain	—
PMMA	Cracks and splinters, difficult to bend	—
Nylon	Springs back	—
Polyvinyl alchol	—	Stretches then pulls apart
PC	Does not bend easily, tough	—
Cellophane	—	Tears easily and straight

8. Burning Test: The observations that may be made with the aid of the flame are:

(*i*) How the material burns in flame?
(*ii*) Whether the material burns after removal from the flame?
(*iii*) Nature of smoke and colour of smoke.
(*iv*) What kind of odour is given off ?
(*v*) The nature of any residue from these observations, the possibilities given in Table 4.3 are considered.
(*vi*) Self-extinguisher or continues to burn.

Table 4.3 gives the behaviour of some common polymers when held in flame for 10 seconds and then withdrawn. To test the behaviour of the polymer in a flame, hold a small sample of the polymer with a pair of tweezers or a spatula in a low flame. Reduce the gas supply to the Bunsen burner to its minimum. Observe the flammability of the plastic in and out of the flame. These polymers are assumed to be uncompounded. However, compounded polymers give different results. For example, highly plasticized PVC may not be self-extinguishing whereas, PMMA containing antimony sulphide may be self-extinguishing. Therefore, in practice results may deviate from those shown in Table 4.3.

9. Pyrolysis Test: Add a small sample (approximately 100 mg) to a pyrolysis tube, gripping the upper end of the tube with a clip or a pair of tongs. At the open end of the tube place a piece of moist litmus or pH paper. Heat the test tube in the flame of a Bunsen burner that has been reduced to minimum heat, taking care to point the open end of the tube away from the face. Heating should occur so slowly that the changes in the sample and in the smell of the decomposition gases can be properly determined. Depending on the reaction of the escaping vapours with litmus paper it is possible to distinguish three different groups: acidic (blue litmus to red), neutral (no colour change), or basic (red litmus changes to blue). Acid vapours may come from carbohydrate polymers and their derivatives. High acid vapours often indicate the presence of chlorine, e.g. PVC, PVDC, chlorinated rubber, etc. Alkaline vapours indicate the presence of nitrogen, e.g. nylons, PU, Urea formaldehyde and melamine-formaldehyde. Neutral vapours are evolved from hydrocarbon polymers, silicone and some polyesters.

Table 4.3: *Burning, melting behavior and odour characteristics of various polymers*

Sl. No.	*Plastic Materials*	*Burning Characteristics and Kind of Smoke*	*Melting Behavior*	*Odour or Smell*
1.	Polyethylene (PE)	Blue base and yellow tip. No smoke, burns continuously.	Melts and drips LDPE-fast HDPE-slow (Neutral fumes).	Waxy smell (extinguished candles).
2.	Polypropylene (PP)	Brilliant with blue centre.	Melts and drips (Neutral fumes)	Lubricating oil smell (extinguished candles).
3.	Polystyrene (PS)	Orange yellow flame, burns continuously with sooty smoke.	Melts but does not drip (neutral fumes).	Mary gold smell (Sweetish).
4.	High impact polystyrene (HIPS)	-do-	-do-	Slight burning, rubber smell.
5.	Acrylo nitrile-butadiene- styrene (ABS)	-do-	-do-	Strong burning, rubber smell.
6.	Styrene-Acrylo nitrile (SAN)	-do-	-do-	Mary gold smell.

Note: Both PS and SAN can be distinguished by elemental analysis.

Sl. No.	*Plastic Materials*	*Burning Characteristics and Kind of Smoke*	*Melting Behavior*	*Odour or Smell*
7.	Polyvinyl alcohol (PVA)	Blue base with yellow tip, burns continuously with bright glowing. Melts but chars around the flame as spirit. No smoke	Melts but chars	Burning sugar smell.
8.	Polyvinyl chloride (PVC)	Self extinguishing	Melts but chars	A white smoke. After the flame is put-off, gives pungent smell of chlorine.
9.	Polyamides Nylon 6, Nylon 66 Nylon 11, Nylon 6,10 Nylon 12	Spurting and self extinguishing	Bubbles and drips fast.	Burning smell like ester or burning hair or wool smell.
10.	Polyester PET PBT	Orange yellow flame with black smoke, burns continuously.	Melts and drips fast.	Fruity smell like ester
11.	Cellulose acetate	Orange yellow flame with black smoke, burns continuously.	Burning surface bubbles (Acidic fumes).	A dense white smoke, smells like vinegar (burned paper).

Note: All polyamides are distinguished from each other by measuring melting point, density and solubility.

12.	Cellulose acetate butyrate (CAB)	—do—	—do—	Butyric smell
13.	Cellulose nitrate	—do—	Explodes	Camphor
14.	Ethyl cellulose	Pale yellow with blue-green base	Acidic fumes	Burning wood
15.	Polymethyl methacrylate (PMMA)	Blue base, yellow tip, burns continuously without smoke.	Burning portion bubbles and does not drip. When the flame input off the melted portion froths.	Fruity smell
16.	Polycarbonate (PC)	Orange yellow flame with black smoke. Self extinguishing.	Melt but chars	Phenolic smell (ink like)
17.	Polyphenylene oxide (PPO)	—do—	—do—	—do—
18.	Polyacetals (POM)	Pale blue flame, burns continuously.	Burning edges boil and slowly drip.	A very pungent smell of formalin.
19.	Polyurethane (PU) MDI based resin TDI based resin	Blue base yellow tip burns continuously, no smoke. Yellow flame self extinguishing black smoke	Drips like oil. Drips very fast.	Faint apple smell or castor oil smell. —do—
20.	Polyphenylene sulphide (PPS)	Orange yellow flame with black smoke, self extinguishing	Melts and chars, very sooty	H_2S smell or rotten egg smell

(contd.)

Sl. No.	Plastic Materials	Burning Characteristics and Kind of Smoke	Melting Behavior	Odour or Smell
21.	Polytetrafluoro ethylene (PTFE)	Does not burn	Becomes transparent when heated strongly and retains opaque on cooling	Evaporation at red heat and piercing odour of HF
22.	Polychloro trifluoro ethylene (PCTFE)	Yellow flame	Strongly Acidic fumes	— Do —
23.	Phenol-formaldehyde (PF)	Self extinguishing Bright sooty	Does not melt, burnt surface swells and cracks (Neutral)	Burning wood (or) coal smell
24.	Urea-formaldehyde (UF)	Self extinguishing	—do—	Dry fish smell
25.	Melamine-formaldehyde (MF)	Self extinguishing Light yellow (white edges)	Alkaline	Ammonia, amine (fishy) and formal - dehyde smell
26.	Epoxies	Yellow orange flame with black smoke, burns continuously	Does not melt	Burns with flouer smell
27.	Unsaturated polyester (USP) resins	—do—	—do—	Ester smell

Identification of Polymers on the basis of Burning Test is given in Flow Chart 4.2.

Flow Chart 4.2 *Burning Test*

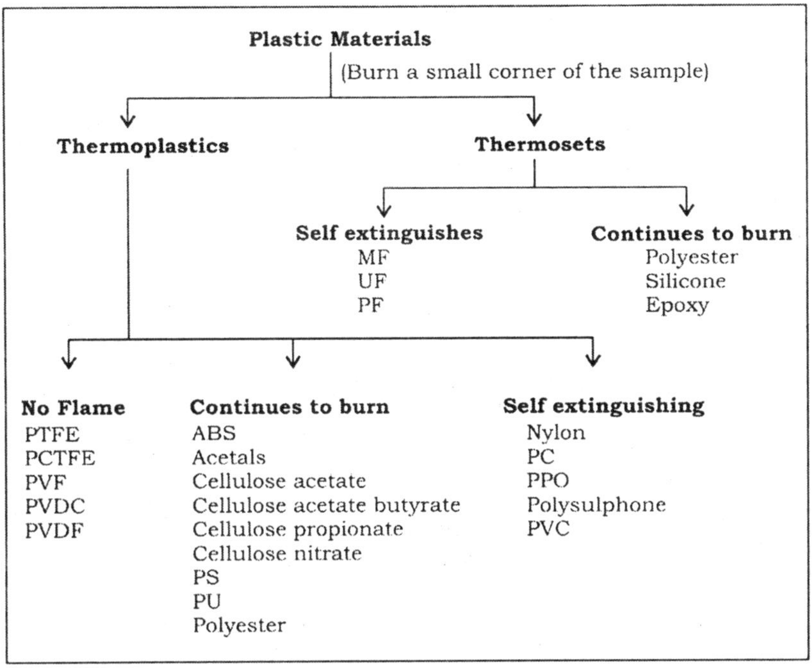

Table 4.4 shows the color reactions of the decomposition products of most important plastics. Depending on their composition, some plastics can appear under different groups in pyrolysis test. For example, PF or PU.

10. Method of fabrication: Relevant information available such as the method of fabrication employed and the end-use to which the material is put, are helpful for identification. Thermoplastics are commonly injection moulded, calendered or extruded. Thermosets are commonly compression moulded and transfer moulded. Thus the sample should be examined for indicative features such as sprue marks, flash, etc. Vacuum-formed articles are likely to be cellulose nitrate, PVC and HIPS.

11. Simple observations: Physical appearances of some polymers can very well distinguish them from other polymers. Some such observations are listed as follows. Also refer to Table 4.2 for some more simple observations.

Table 4.4: *Litmus tests for vapours of plastics*

Red (pH 0-5)	*Unchanged (pH 5.0–5.5)*	*Blue (pH 8.0-9.5)*
Halogen containing polymers (PVC, PVDC), polyvinyl esters, cellulose esters, PET, PBT, PF, PU elastomers, unsaturated polyester resin, fluorine containing polymers, vulcanized fibre, poly alkylene sulphide	polyolefines, polyvinyl alcohols, polyvinyl acetals, polyvinyl ethers, styrene polymer (included styrene-acrylonitrile copolymer)*, PMA, POM, PC, linear PU, silicones, phenolic resins, epoxy resins, crosslinked PU	PAN, ABS, amino resins, phenolic and cresol resins, polyamides

*Some samples show slightly alkaline behaviour.

The pH values of vapour for some high temperature resistant thermoplastics are given in Table 4.5.

Table 4.5: *pH values of pyrolysis vapours from high temperature resistant thermoplastics*

Acidic (pH 0-5)	*Neutral (pH 6-7)*	*Basic (pH 8-12)*
PAR	PEEK	PEI
PESU	PEEK (amorphous)	PA 6-3-T (polytrimethyl hexamethylene terephthalate)
PPS	—	—

⇒ Polyolefines have a waxy feel–LDPE, HDPE and PP

⇒ PS, PC, PMMA, CSA, PET and PBT are transparent if unfilled.

⇒ Teflon (PTFE) feels greasy and is milky white in pure form.

⇒ Unfilled PS moldings give a metallic ring sound when dropped on a hard surface.

⇒ CSA is a bit softer than PMMA and can more easily be marked with the nail.

⇒ Cellulose nitrate cannot be injection molded.

Note: **LDPE** - Granules are chewing in nature, films are soft and translucent
HDPE - Hard and translucent
PP - Hard, glossy, films making noise after rubbing, clear transparent.

II. SOLUBILITY TEST

Solubility tests form a basis for the identification of some main types of plastics. However, in many cases solubility varies considerably for different samples of the same resin and it is difficult to interpret the results. Table 4.6 gives a list of solvents and non-solvents for some plastics. Usually polar polymers are soluble in polar solvents and non-polar polymers are soluble in non-polar solvents. The solubility test should be carried out in a test tube. Take about 100 mg of the sample (granule or powder) and add about 10 ml of solvent, mix and occasionally shake the contents of the test tube and keep it on water bath. Observe for few minutes. Swelling or sticking to the wall of test tube or dissolution of polymer indicates that the polymer is soluble in that solvent.

The solubility of a polymer material depends very much on its chemical structure and to some extent on its molecular weight. The solvents mentioned in Table 4.6, therefore, donot always permit an unambiguous identification.

***Table 4.6:** Selected solvents and non-solvents for some polymers*

Polymer	*Solvents*	*Non solvents*
Cellulose acetate	Methylene chloride, chloroform, trichloroethane, tetrahydrofuran, dioxane, acetone, ethyl acetate	Aliphatic hydrocarbons, benzene, dichloro ethane, chloro benzene, ethanol
Cellulose nitrate	Halogenated hydrocarbons, acetone, methyl ethyl ketone, cyclo hexanonė, ethyl acetate, ethyl butyrate, ethylene glycol, ethylene acetate, nitrobenzene	Aliphatic hydrocarbons, aromatic hydrocarbons, lower alcohols, ethylene glycol, diethyl ether, water

Polymer	*Solvents*	*Nonsolvents*
Ethyl cellulose	Benzene, toluene, methylene chloride, alcohols, esters	Hydrocarbons, dekaline, xylene, carbon tetrachloride, n-propyl ether
Cellulose esters, methyl cellulose	Methanol, methylene chloride, formic acid, acetic acid	Hydrocarbon, decalin, xylene
Cellulose acetate butyrate/ propionate	Acetone, benzene, methylene chloride	—
Cellulose propionate, cellulose nitrate	Chlorinated hydrocarbons, benzene, lower alcohols, ketones, acetic esters, acetic acid	—
Carboxy methyl cellulose	Water	Methanol
Nylon 6,6	M-cresol, chloro phenol, formic acid, acetic acid, phosphoric acid	Hydrocarbons, chloroform, alcohols, ethers, ketones, and esters
Nylon 6	Trifluoro ethanol, trichloro ethanol, phenol, cresols, formic acid, halogenated acetic acid, sulphuric acid, phosphoric acid	Hydrocarbons, aliphatic alcohols, chloroform, diethyl ether, aliphatic ketones and ethers
Polyacrylonitrile	0-, m-, and p-phenylene diamines, n-nitrous piperidine, maleic anhydride, dioxanone, ethylene oxalate, propylene carbonate, 2-oxazolidone, dimethyl formamide, cyano acetic acid	Aliphatic/aromatic hydrocarbons, chlorinated hydrocarbons, alcohols, diethyl ether, ketones, acrylonitrile

Polymer	*Solvents*	*Nonsolvents*
Polybutadiene/ Polyisoprence/ Natural rubber	Aliphatic and cycloaliphatic hydrocarbons, tetrahydrofuran, higher ketones, higher aliphatic ester	Alcohol, lower ketones and ester, nitromethane, propionitrile, water
Polybutyl acrylate	Aromatic hydrocarbons, chlorinated hydrocarbons, tetrahydrofuran, ester, ketones, butanol	Aliphatic hydrocarbons, diethyl ether, methanol, ethanol, ethyl acetate
Polyethyl acrylate	Aromatic hydrocarbons, chlorinated hydrocarbons, tetrahydrofuran, esters, ketones, methanol, ethanol, butanol, glycol, ether	Aliphatic alcohols of more than 5 C-atoms, cyclohexanol, aliphatic hydrocarbons, diethyl ether
Polyethylene (i) high density	Above 80°C; aliphatic, cyclo aliphatic and aromatic hydrocarbons, higher aliphatic esters and ketones, di-n-aryl ether, dekaline.	All common organic solvents at room temperature, highly polar organic solvents at elevated temperatures
(ii) Low density	As above but dissolve between 40° and 60°C	As above
Polyisobutylene	Aliphatic, cycloaliphatic, and aromatic hydrocarbons, chlorinated hydrocarbons	Lower ketones and esters, lower alcohols, nitromethane, cyclopropionitrile
Polymethyl acrylate (PMA)	Aromatic hydrocarbons, chlorinated hydrocarbons, tetrahydrofuran, ester, ketones	Aliphatic hydrocarbons, diethyl ether, methanol, ethanol, carbon tetrachloride
Polymethyl methacrylate (PMMA)	Benzene, toluene, xylene, methylene chloride, chloroform, chloro benzene, dioxane, methyl ethyl ketone, cyclohexane, acetic acid, ethyl acetate	Hexane, cyclohexane, mineral oils, methanol, ethylene glycol and glycerol

Polymer	*Solvents*	*Nonsolvents*
Polypropylene (i) Atactic (ii) Isotactic	Hydrocarbons and chlorinated hydrocarbons, diethyl ether Same as polyethylene	More polar organic solvents with a small hydrocarbon groups, (at elevated temperature)
Polystyrene/ ABS, SAN	Cyclohexane (above 35°C), benzene, toluene, ethyl benzene, lower chlorinated hydrocarbons, tetrahydrofuran, dioxane, methyl ethyl ketone, diisopropyl ketone, cyclohexanone, ethyl acetate, carbon disulfide	Saturated hydrocarbons, alcohols, diols, ethylene chlorohydrine, diethyl ether acetate
Polycarbonate (PC)	Cyclohexanone, dimethyl formamide, p-cresol, methylene chloride	—
Polyacetal	Dimethyl formamide (DMF) (150°C), dimethyl sulphoxide	—
Polyurethane (PU)	Dimethyl formamide, tetrahydrofuran, formic acid, ethyl acetate, m-cresol	Methanol, diethylether, hydrocarbons
Polyvinyl acetate/ Polyvinyl butyrate	Benzene, toluene, chloroform, chlorobezene, methanol, allyl alcohol, benzyl alcohol	Saturated hydrocarbons, ethanol, cyclohexanol, diethyl ether, carbon disulphide
Polyvinyl formal	Ethylene dichloride, dioxane, glacial acetic acid, phenol	—
Polyvinyl alkyl ethers/ polyvinyl carbazole	Chlorinated hydrocarbons, esters, benzene, alcohols, water, ketones	—

Polymer	*Solvents*	*Non solvents*
Polyvinyl alcohol (PVA)	Glycol (hot), glycerol (hot), tri ethylene diamine formamide, dimethyl formamide, dimethyl sulphoxide (hot), water	Hydrocarbons, chlorinated hydrocarbons, lower alcohols, tetra-hydrofuran, dioxane, ketone, ester
Polyacrylamide	Water	Methanol, acetone
Polyacrylic acid	Water, dilute alkalis, methanol, dioxane, DMF	Hydrocarbons, acetone, diethyl ether
Polyvinyl chloride (PVC)	Tetrahydrofuran, methyl ethyl ketone, cyclopentanone, cyclohexane, dimethyl formamide, nitrobenzene, dimethyl sulphoxide	Aliphatic hydro-carbons, mineral oils, aromatic hydrocarbons, alcohol, vinyl chloride, ester, carbon disulphide
Polyvinylidene chloride (PVDC)	Tetrahydrofuran, butyl acetate (hot), dimethyl formamide, ketones	—
Polytetrafluoro ethylene (PTFE)	Hot fluorocarbon oil	—
Polyvinyl fluoride/ Polyvinylidene fluoride	Above 110°C; cyclo hexanone, dimethyl formamide, dimethyl sulphoxide	Aliphatic hydro-carbons, methanol
Alkyd resin		—
(i) Uncured	Chlorinated hydrocarbons, lower alcohols, esters, benzyl amine (160°C), ammonia	
(ii) Cured	Insoluble	
Epoxy resin		
Uncured	Alcohols, dioxane, ketones, ester	
Cured	Insoluble	—

Polymer	*Solvents*	*Non solvents*
PF, UF, MF		
Uncured	Alcohols, ketones, benzyl amine (200°C), hot alkalis	
Cured	Insoluble	—
Aliphatic polyesters	Chloroform, formic acid, benzene	Methanol, diethylether, aliphatic hydrocarbons
Polyethylene terephthalate (PET)	m-cresol, o-chloro phenol, nitrobenzene trichloroacetic acid	Methanol, acetone, aliphatic hydrocarbons
Polyoxymethylene (POM)	Benzyl alcohol, dimethyl formamide	Methanol, diethyl ether, aliphatic hydrocarbons
Poly ethylene oxide	Water, benzene, dimethyl formamide	Diethyl ether, aliphatic hydrocarbons

III. ELEMENTAL ANALYSIS

Depending upon the constituent atoms, polymers are classified as organic and inorganic. Organic polymers essentially contain carbon and hydrogen. Organic polymers are further classified as homochain and heterochain polymers. Homochain polymers contain only carbon and hydrogen, whereas heterochain polymers contain heteroatoms like oxygen, nitrogen, sulphur, etc., apart from carbon and hydrogen. Inorganic polymers do not contain any carbon. Hence, elemental analysis can very well distinguish some polymers.

Tests for Heteroatoms

The elements that are present in the plastic in addition to carbon and hydrogen, are nitrogen, sulphur, chlorine, fluorine, silicone and sometimes phosphorous. Unfortunately there is no simple direct method for the certain identification of the oxygen, so it is not possible to test for oxygen in a qualitative manner. A compilation of the most important plastics containing heteroatoms is shown in the Table 4.7.

Table 4.7: *Classification of plastics based on their heteroatoms*

					Heteroatoms				
(Nil)	*O*		*Halogens*	*N, O*	*S, O*	*Si*	*N, S*	*N, S, P*	*Polymers not containing N, S, halogens*
	Cannot be saponified	Can be saponified							
Polyolefines	Polyvinyl alcohol	Polyvinyl acetate & co-polymers	PVC	Poly amides	Poly alkylene sulphide	Silicones	Thiourea conden-sates	Casein resins	Cellulosics
PS	Polyvinyl ethers	Poly acrylates & PMA	PVDC & copoly-mers	PUs	Vulcan-ized rubber	Poly siloxanes (PDMS)	Sulphamide condensates		Epoxides
Poly-isoprene	Polyvinyl acetals	Polyesters	Poly fluro hydro-carbons	Poly urea	Poly ether sulfone	Silicone-epoxy			PF, PC, PE, PP
Butyl rubber	Poly glycols	Alkyd resins	Chlorinated rubbers	Amino-plastics	Poly aryl sulfone	Fluoro silicone rubber			PET, PVAc

(*contd.*)

			Heteroatoms					
(Nil)	*O*	*Halogens*	*N, O*	*S, O*	*Si*	*N, S*	*N, S, P*	*Polymers not containing N, S, halogens*
Poly aldehydes	Cellulose esters	Polytetra fluoro ethylene	PAN & copolymers	Polysulfone				Polyformal-dehyde
Phenolic resins		Polychlorotri-fluoro ethylene	Polyvinyl carbozole					Shellac, PVA
Cellulose		Polyvinyl fluoride	Polyvinyl pyrrolidone					Alkyds
Cellulose ethers, Xylene resins		Polyvinyl bromide	UF, MF, cellulose nitrate					

Preparation of Sodium Fusion Extract

A small piece of clean sodium (about 0.02 g) is placed in an ignition tube, warmed and about 0.1 g of the sample to be tested is added and covered with little Na_2CO_3. The tube is gently heated until the reactions subside and then strongly heated until the tube is red hot. The tube is then plunged into 10ml of distilled water contained in a pestle and mortar. The tube is crushed and the resulting solution is boiled and filtered. The filtrate is divided into 4 parts.

(a) Test for Nitrogen: The first portion of the sodium fusion extract is boiled with 3 drops of freshly prepared aqueous ferrous ammonium sulphate (approx. 5%) solution and cooled. This is acidified with 5N sulphuric acid and a drop of 0.5N aqueous ferric chloride solution is added. A substantial blue precipitate of ferric isocyanates (Prussian blue) indicates the presence of nitrogen. Presence of blue colour on the filter paper is also to be taken as positive result. A slight blue precipitate can be taken as negative result. Nitrogen containing plastics include polyamides, PU, MF, UF, SAN, ABS, etc.

(b) Test for Halogens

(i) Test for Chlorine: Sodium fusion extract part is acidified with 5N nitric acid and boiled for 2 minutes. 0.1 N aqueous silver nitrate is added to this. A white precipitate soluble in ammonia, confirms the presence of chlorine. Chlorine containing plastics include PVC, CTFE, PVDC, PCTFE, etc.

(ii) Test for Bromine: Sodium fusion extract part is acidified with 5N nitric acid and boiled for 2 minutes. 0.1 N aqueous silver nitrate is added to this. A yellow precipitate insoluble in ammonia, confirms the presence of bromine, e.g. polyvinyl bromide (PVB).

(iii) Test for Fluorine: An aqueous solution of zirconium nitrate (0.1%) in alizarin red (0.1%) is prepared. Filter paper is immersed in this solution and then allowed to dry. When required, a small piece of this filter paper is moistened with aqueous acetic acid (50%). The sodium extract is neutralized with 5N hydrochloric acid and a drop of this is placed on the moistened filter paper. A red spot turns yellow indicating the presence of fluorine. Fluorine containing polymers include PVF, PVDF, PCTFE, PTFE, etc.

(c) Test for Sulphur: 3 drops of freshly prepared aqueous sodium nitro prusside (approx. 5%) are added to the sodium fusion extract. A violet colour indicates the presence of sulphur. A transient coloration only indicates small amount of sulphur. Sulphur containing plastics include PPS, polysulfone, polythiophene, etc.

(d) Test for Carbon

1. About 0.05 g of sample is mixed with 0.02 g of potassium dichromate and 10 drops of phosphoric acid. The mixture is heated under vacuum in a test tube on a glycerol or sulphuric acid bath at 200°C. The test tube is connected to an U-tube containing clear barium chloride solution. White precipitate of barium carbonate confirms the presence of carbon.

2. About 2.5 g of sample is mixed with equal amount of each sodium and ammonium sulfates. The mixture is heated until it glows red. After cooling, the mixture is dissolved in 3 ml water and boiled with 2 mg of ferrous sulphate and acidified with 10% hydrochloric acid. A blue colour indicates the presence of carbon, whereas yellow, brown or green colour indicates the absence of carbon. Inorganic polymers do not contain carbon.

(e) Test for Oxygen: 1 g of ferric chloride and 1 g of potassium thiocyanate are dissolved separately in 10 ml of methanol. Then two solutions are mixed. After standing for few hours, the potassium chloride precipitate is filtered off. The filter paper strips are dipped in the above solution and dried in air. A few drops of liquid polymer are placed on this filter paper. Deep wine red colour indicates the presence of oxygen. Oxygen containing polymers include PC, PET, EVA, EVOH, PU, PMMA, PEA, PPO, PF, UF, MF, POM, epoxy, polyamides, etc.

(f) Test for Hydrogen: Sample and 1 mg of sulfur are placed in a test tube and heated to 220-250°C. The mouth of test tube is covered with a lead acetate paper. A brown spot on the paper indicates the presence of hydrogen. Almost all polymers contain hydrogen except PTFE, CTFE and silicones.

(g) Test for Phosphorus: 1 g of sample is heated with 3 ml of concentrated nitric acid and few drops of concentrated sulphuric acid, cooled, diluted with water and some drops of molybdate solution (preparation given below) is added and heated. A yellow precipitate indicates the presence of phosphorus. Phosphorus containing polymers include phosphanes, phosphonitrile, etc.

Preparation of Molybdate Solution

Dissolve 30 g ammonium molybdate in hot water, cool and add 100 ml water. Pour this solution slowly into a solution of 10 g of ammonium sulphate in 10 ml of 55% nitric acid and mix very well. Let this stand for one day and filter through a thick acid proof filter paper and store it in well sealed container in dark.

IV. TRANSITION TEMPERATURES

Glass transition temperature (T_g) and crystalline melting temperature (T_m) are also decisive factors, which can distinguish between different polymers and polymer blends. The temperature below which a polymer is hard and above which it is soft, is called the ***glass transition temperature***. The hard, brittle state is known as the glassy state and the soft flexible state is known as the rubbery or viscoelastic state. Glass transition temperature is the approximate mid point of the temperature range over which the glass transition takes place.

A cystalline substance can thus exist as a crystalline solid (solid phase state) or as a molten liquid (liquid phase state), which solid-to-liquid transition occuring at a temperature known as the ***crystalline melting point*** (T_m). T_m exists only for thermoplastics or crystalline polymers. Thermosets are characterized by their degradation temperature and T_g. Elastomers are characterized only by T_g. T_m is determined by Koffler's apparatus. T_g and T_m can be determined by thermal methods like TMA and DSC. Information about transition temperatures is given in Table 4.1.

V. FINAL IDENTIFICATION BY CHEMICAL TESTS

The confirmatory tests by chemical methods for few common polymers are briefly described below:

General Identification Reactions

The reaction described here are useful as screening tests for certain groups of plastics, but also for testing for specific cleavage products, such as phenols or formaldehyde, from some plastics.

Liebermann-Storch-Morawski Reaction

Dissolve or suspend a few milligrams of the sample in 2 ml hot acetic anhydride. After cooling, add 3 drops of 50%

sulphuric acid (from equal volumes of water and concentrated sulphuric acid). Watch the colour reaction immediately and again after the sample has stood for 10 min. Heat it to 100°C, using a water bath. This test is not specific, but is often quite useful as an indicator. The colour changes after treating with the reagents are given in Table 4.8.

Table 4.8: *Colour changes in the L-S-M reaction*

Material	*Immediate colour change*	*Colour after 10 min.*	*After heating to about 100°C*
Phenolic Resins	Reddish violet-pink	Brown	Brown-red
Polyvinyl alcohol	Colourless-yellowish	Colourless-yellowish	Brown-black
Polyvinyl acetate	Colourless-yellowish	Blue-grey	Brown-black
Chlorinated rubber	Yellow-brown	Yellow-brown	Reddish to yellow-brown
Epoxy resins	Colourless to yellow	Colourless to yellow	Colourless to yellow
Polyurethanes	Lemon yellow	Lemon yellow	Brown, green fluorescence

SPECIFIC IDENTIFICATION TESTS

Group 1: Plastics Containing Nitrogen

1. Test for ABS

1 g of polymer is refluxed with 20 ml of concentrated nitric acid for 1 hour, using long air condenser in a fuming cupboard. The mixture is poured into 100 ml of water and extracted with 50 ml of ether. Ethereal extracts are washed with water. Aqueous extracts are rejected and the rest of the extract is acidified with 20 ml of hydrochloric acid. 5 g of granulated zinc is added to this and the mixture is heated on a steam bath for 20 minutes. The mixture is cooled below 5°C in a freezing mixture and freshly prepared sodium nitrite (0.1 g in 2 ml of water) is added. The resulting solution is poured into freshly

prepared solution of 0.05g 2-naphthol in 10 ml of 5N aqueous sodium hydroxide. A vivid scarlet colour confirms the presence of ABS.

2. Test for Polyamides

(i) The sample is heated in formic acid. The following observations are made:

Soluble at room temperature	⇒	Nylon 6 or Nylon 6,6
Soluble in boiling formic acid	⇒	Nylon 6,10
Insoluble in formic acid (cold and hot)	⇒	Nylon 11 or Nylon 4,6

(ii) 0.2 g of the sample is heated on a low flame in an ignition tube and the vapour coming out is passed through a filter paper soaked in a saturated solution of o-nitro benzaldehyde in 2N aqueous sodium hydroxide. A deep violet colour confirms the presence of nylon 6,6 (adipic acid).

(iii) About 0.5 g of the sample is heated in 5 ml distilled water and allowed to boil for 10–15 minutes. After cooling, 2–3 drops of concentrated sulphuric acid is added to 0.5 ml of the above solution followed by the addition of 2 ml of potassium iodo-bismuthate (a solution of 5 g of basic bismuth nitrate and 25 g of potassium iodide in 10 ml of 2% sulphuric acid. Precipitation of an orange red complex indicates the presence of poly caprolactum (nylon 6).

(iv) Different polyamides can also be distinguished by their nitrogen content. Nitrogen content of different polyamides are: Nylon-6 (12.4%), Nylon-6,6 (12.4%), Nylon-11 (7.6%), and Nylon-6,10 (9.9%).

(v) Polyamides can be readily identified by the acids formed on acid hydrolysis of the respective polyamides. For this purpose, heat 5 g of the sample with 50 ml conc. HCl using a reflux condenser. Continue reflux until the major part of the sample has dissolved. Then boil the solution with charcoal until the color disappears and filter it while it is hot. After cooling, filter of the precipitated acids and recrystallize them from a small amount of water. If no acids precipitate extract the filtrate with ether. Evaporate the ether and recrystallize the residue from water. The acids have the following melting points.

Adipic acid (Nylon 6, 6)	152°C
Sebacic acid (Nylon 6, 10)	133°C
ω-Aminocaproic acid hydrochloride (Nylon 6)	123°C
11–Amino undecanoic acid (Nylon 11)	145°C
12-Amino auric acid (Nylon 12)	163°C

Polyamides are easily differentiated from polyacrylonitrite (PAN) by dissolution of the sample in dimethyl formamide (DMF) and subsequent addition of NaOH solution.

3. Test for Polyurethanes

(i) About 0.5 g of the sample is dissolved in 10 ml of glacial acetic acid. Then, 0.1 g of di methyl amino benzaldehyde is added. The solution turning yellow after several minutes shows the presence of polyurethane.

(ii) Reaction with p-nitrophenyl diazonium fluoroborate

A filter paper is moistened with 1% of p-nitro-phenyl diazonium fluoroborate. The filter paper is inserted in the fumes produced from the pyrolysing of the sample. The di-isocyanate used in the manufacture of polyurethane (PU) can be identified according to the colour that develops in the paper as given below;

Reddish brown	⇒	Toluene di-isocyanate (TDI)
Violet	⇒	naphthalene di-isocyanate (NDI)
Yellow	⇒	methylene di-isocyanate (MDI)
Beige	⇒	phenylene di-isocyanate (PDI)

(iii) Colour reaction to differentiate polyester-urethanes from polyether urethanes

A sample of about 5 g is finely powdered and treated with a few drops of sodium hydroxide solution (2N methanolic solution with little phenophthalene) and then with a few drops of a saturated methanolic solution of hydroxyl amine hydrochloride solution. The mixture must be alkaline. For highly crosslinked polymer, the mixture is heated for 20-30 sec. The mixture is then acidified with 1N hydrochloric acid and each drop of the acid is followed by an addition of 3% ferric chloride reagent until a colour develops.

In the presence of a polyester-urethane, the mixture turns violet. In the case of polyester-urethane based on castor oil or dimerised fatty acid, the mixture turns brown or violet brown. Polyether-urethanes do not give the characteristic colour reaction.

4. Tests for Poly Acrylonitrile (PAN)

(i) The sample is dissolved in dimethyl formamide, made alkaline with concentrated sodium hydroxide solution and heated, development of orange red colour indicates the presence of polyacrylonitrile.

(ii) A small amount of zinc dust and a few drops of 25% sulphuric acid are added to the sample. The mixture is heated in a porcelain crucible. The below given reagent is added to this mixture. PAN develops a blue colour.

Reagent preparation: Dissolve 2.86 g of copper acetate in one litre of water. Then dissolve 14 g of benzidene in 100 ml of acetic acid and to 67.5 ml of this solution add 52.5 ml of water. Keep both the solutions in dark. Mix them in equal volumes just before use.

(iii) The sample is heated with 20% hydrochloric acid. The insoluble polyacrylic acid precipitates and confirms the presence of PAN

5. Test for Polyvinyl Pyrrolidine (PVP)

0.5 g of the sample is dissolved in 10 ml of water and 1 ml of 0.01 N aqueous iodine is added. A blue colour confirms polyvinyl pyrrolidine (PVP).

6. Test for Cellulose Nitrate (CN)

(i) Dissolve the sample in acetone, add 2- 3 drops of 2% ethanolic solution of 2-napthanol and slowly add concentrated sulphuric acid through the sides of the test tube. A green ring at the junction of two solutions indicates the presence of cellulose nitrate.

(ii) *Diphenyl amine test:* Boil the sample with 0.5 N aqueous sodium hydroxide or potassium hydroxide, cool, acidify with dilute sulphuric acid and filter. Coagulate the lower layer in the filtrate by adding diphenyl amine (10 g diphenyl amine in 3ml water and 10 ml concentrated sulphuric acid) solution and shake. Blue ring at the interface confirms cellulose nitrate.

(iii) Add 1 ml of water to a test tube containing about 5 mg of the sample. Add 2 ml of freshly prepared 0.2% solution of anthrone in concentrated sulphuric acid. A green colour changing to blue confirms the presence of cellulose nitrate.

7. Test for Cellulosics

(i) *Test for acetates and propionates:* The sample (about 0.2 g) and few ml of hydrochloric acid are heated for 10 minutes and dilute ammonia is added to make alkaline. One ml of this solution is reacted with 1-2 drops of a 5% lanthanum nitrate solution and 1 drop of 0.1 N iodine solution. A fast colour change to blue confirms the presence of acetates and a brown colour indicates the presence of propionate. For differentiation between cellulose acetate and cellulose acetate-butyrate, it is usually sufficient to examine the vapour produced by dry heating of the sample. The acetate smells like acetic acid, the acetate-butyrate has the smell of both acetic acid and butyric acid (like rancid butter).

(ii) *Test for cellulose ethers*: The important cellulose ethers are methyl cellulose, ethyl cellulose and benzyl cellulose.

Reaction to heating and burning—Cellulose ethers melt and char after heating and continuously burn after ignition. Nature of flame and smell of the cellulose ethers are as follows:

Methyl cellulose	–	Brilliant yellow flame
Ethyl cellulose	–	Less brilliant flame accompanied by dripping
Benzyl cellulose	–	Brilliant sooty flame smells like bitter almonds

Note: All these ethers will have the odour of burnt paper.

(iii) *Test for methyl cellulose:* The sample is treated with a 10% tannic acid solution. A flaky (flocculent) precipitate confirms the presence of methyl cellulose.

The sample is treated with iodine – potassium iodide solution. A colour change from violet brown to brown, which disappears on the addition of strong NaOH confirms the presence of methyl cellulose.

(iv) *Test for ethyl cellulose:* The sample is treated with one drop of dichromate solution (1.0 g of potassium dichromate $K_2Cr_2O_7$ + 60 ml water + 7.5 ml of conc. H_2SO_4) and then heated to 100°C. The mouth of the tube is covered with a filter paper moistened with a mixture of equal volumes of 20% aqueous morpholine solution and 5% aqueous sodium nitroprusside. A blue colour indicates the presence of ethyl cellulose.

(v) *Molisch test:* Sample is dissolved in acetone, reacted with 2 to 3 drops of 2% ethanolic solution of 1-napthanol and a few drops of concentrated sulphuric acid is allowed to flow down along the wall of the test tube. Red to reddish brown ring at the interface indicates cellulose.

(vi) *Aniline-acetate reaction:* Around 10mg of sample and a drop of syrupy phosphoric acid is heated in a micro crucible to 40°C. The mouth of the crucible is covered with a filter paper moistened with aniline-acetate solution (10% solution of aniline in 10% acetic acid). The bottom of the crucible is cautiously heated for 30-60 sec. A pink or red stain indicates the presence of cellulose derivatives.

8. Test for Melamine-Formaldehyde (MF)

(a) *Test for Melamine*

(i) About 0.5 g of the sample is refluxed with 80% acetic acid (25 ml) for 30 min. The mixture is cooled and filtered. The filterate is evaporated to dryness in a small evaporating dish on a steam bath. Add 2 ml of water, in a small evaporating dish on a steam bath. Add 2 ml of water, cooled and filtered. To the filtrate 1 drop of saturated aqueous picric acid is added. A yellow precipitate indicates the presence of melamine.

(ii) *Thiosulphate reaction:* The sample is heated with few drops of conc. hydrochloric acid in an ignition tube at 190-200°C in a glycerin bath, until litmus paper no longer turns blue when placed in the mouth of the test tube. A few crystals of sodium thiosulphate are added to cold residue, the tube is covered with a piece of congo paper moistened with a paper moistened with 3%

hydrogen peroxide and the tube is heated to 160 °C in a glycerin bath. A blue colour on the paper confirms the presence of melamine.

(b) *Test for formaldehyde:* To 0.01 g of the sample, 1.5 ml of conc. sulphuric acid and 1 ml water are added. To this solution few crystals of chromotropic acid is added and the mixture is heated for 10 min on water bath at 65-75°C. A violet/dark violet colour indicates the presence of formaldehyde.

9. Test for Urea-Formaldehyde (UF)

(a) *Test for Urea*

(i) About 0.05 g of sample is refluxed for 10 min with 25 ml of 20% acetic acid. The mixture is cooled and filtered, to the filtrate, 2 ml of xanthydrol in methanol (1%) is added and the mixture is boiled for 1-2 min. A white bulk, precipitate confirms the presence of urea.

(ii) A small quantity of the sample is heated with 14% potassium hydroxide solution in ethylene glycol. Since on hydrolysis, urea decomposes into carbon dioxide and ammonia, the evolving ammonia can be identified by its characteristic smell or by means of a wet red litmus paper turns blue confirms the presence of urea.

(iii) A few milligram of the polymer sample is treated with one drop of concentrated hydrochloric acid and evaporated to dryness at 100°C. The residue is cooled, treated with one drop of phenyl hydrazine hydrochloride and heated for 5 min on an oil bath at 195°C. After being cooled, the product is mixed with 3 drops of 1:1 ammonia solution and 5 drops of 10% aqueous nickel sulphate solution and shaken with 10-12 drops of chloroform. A colour in the chloroform layer indicates the presence of urea resin.

(b) Detection and Differentiation of Urea and Melamine Eesins

(i) To about 1 g of the sample, 15 g of p-dimethyl amino benzaldehyde and 10 ml of mixture of 10 parts acetic anhydride and 3 parts acetic acid in a 50 ml round bottomed flask and boil under reflux for 10 min.
Urea resin gives blue to blue-green solution and greenish colour for just a small amount of urea resins.
Melamine resin gives colorless precipitate at the condensation point in the reflux condenser.

(ii) A few milligrams of the test sample are treated with one drop of concentrated hydrochloric acid and evaporated to dryness at 100 °C. The residue is cooled, treated with one drop of phenyl hydrazine hydrochloride and heated for 5 min on an oil bath at 195 °C. After being cooled, the product is mixed with 3 drops of 1:1 ammonia solution and 5 drops of 10% aqueous nickel sulphate solution and shaken with 10-12 drops of chloroform. A violet colour in the chloroform layer indicates the presence of urea resin.

(iii) The sample is heated to boiling for short time with 3 ml concentrated hydrochloric acid and the solution is divided into two parts. One part of the solution, is made alkaline with dilute sodium hydroxide and sodium hypochlorite. Second part of the solution, is treated with freshly prepared furfural reagent (5 drops of pure freshly distilled furfural + 2 ml of acetone + 1 ml of conc hydrochloric acid + 2 ml water).

Solution 1 remains colourless and solution 2 becomes orange to red after 3-5 hours. A small evolution of heat may occur. This confirms the presence of urea resin. Solution 1 slowly becomes orange and solution 2 remains colorless. This indicates the presence of melamine resin.

Group 2: Plastics Containing Chlorine

10. Test for Chlorine Containing Polymer

(i) Heat a small amount of polymer in pyridine and make a solution. Add a few drops of 5% methanolic sodium hydroxide to a portion of boiling solution. Keep another portion of the solution and let it be cool. Observe the colour change and refer Table 4.9.

(ii) *Beilstein Test (copper wire test)*: A bright copper wire (about 5 cm long with one end is embedded in cork, which serve as a handle) is cleaned by heating to redness in a colourless Bunsen flame. The sample under test is touched by the hot wire, which is returned to the flame. A distinct green or green-blue colour indicates the presence of halogens (chlorine, bromine, iodine and fluorine). It should be noted that a compounding ingredient may contain halogens and thus gives rise to a positive result.

Table 4.9: *Colour test to identify chlorine containing polymer. Polymers after treatment with pyridine and alcoholic NaOH*

Polymer	*In Boiling Solution*		*In Cooled Solution*	
	Immediate	*After 5 min*	*Immediate*	*After 5 min*
Polyvinyl chloride	Olive green	Reddish brown	Yellowish	Olive green
Chlorinated polyvinyl chloride	Blood red	Blood red	Brown	Dark brown
Polyvinyl-idene chloride (PVDC)	Blood red	Brown black	Precipitate in brown colour	Black-brown ppt.
Polyvinyl chloride-vinyl acetate copolymers	Yellow brown	Brown	Colorless	Bright yellow

(iii) *Colour reaction with chloro acetic acid and dichloro acetic acid:* Melt approximately 5 ml mono chloro acetic acid and/or 5 ml dichloroacetic acid, mix with a few milligrams of the sample and heat to boiling for 1-2 min. From the colour that occurs, it is possible to identify the polymers present. If no colour is produced after 2 minute of boiling, the test is negative. If colour is produced after 2 minutes, refer Table 4.10.

Table 4.10: *Colour test for chlorine analysis*

Polymer	*Reaction with monochloro acetic acid*	*Reaction with dichloro acetic acid*
Polyvinyl acetate	Red-violet	Red-violet
Polyvinyl chloride	Bright green	Red-violet
Polyvinyl carbazole	Bright green	Blue
Polyvinyl ethers		
Methyl	Green	Blue-violet
Ethyl	Blue-green	Green-blue
Polyvinyl pyrrolidine	Rose violet, very fast change to blue green	—
Vinyl (chloride-vinyl acetate) copolymers	Wine red to violet	Blue to violet

(iv) *Humal Colour Reaction:* Suitable tests for distinguish between chlorinated polymers are briefly explained below.

Preparation: Free the sample from plasticizers, by extracting with other. Report the Beilstein test, for chlorine to verify the presence of a chlorinated plastic. Dissolve the sample in tetrahydrofuran, centrifuge or filter the solution and precipitate the polymer product again by mixing with methanol. Filter and dry at a maximum temperature of 80°C.

Colour Reaction with Cold Pyridine: The plasticizer free sample is treated with 1 ml pyridine and allowed to stand for few minutes and then add 2-3 drops of approximately 5% methanolic sodium hydroxide. The developing colour is observed immediately and after about 5 min and again after 1 hour. The plastic can be identified by verifying with Table 4.11.

(v) *Reaction with morpholine:* A small amount of sample (0.5 g) is dissolved in morpholine by heating on a water bath. The developing colour is observed immediately and after some time (Table 4.12).

Chlorine percentage in various chlorinated polymers: Percentage of chlorine in chlorinated polymers serves as an identification to distinguish between them. Table 4.13 gives percent chlorine content in some polymers.

11. Test for PVC

Gently heat 0.5g of sample in a test tube with 10 ml of water and 0.01 N aqueous iodine. A development of blue colour confirms the presence of PVC.

Group 3: Plastics Containing Fluorine

Of a large number of possible fluorine containing polymers are: polytetrafluoro-ethylene (PTFE), polyvinyl fluoride (PVF), polychlorotrifluoroethylene (PCTFE), and poly-vinylidene fluoride (PVDF), which have been used in the most of plastics industry. PTFE and PCTFE are used especially for parts that require high heat resistant. PVF and PVDF are extraordinarily resistant to chemicals, solvents and ether. They are used in the production of films and protective coating for metals, plastics and woods.

Table 4.11: *Colour reactions with cold pyridine of chlorinated polymers (Humal Colour Reaction)*

Materials	*Boiled with pyridine and added reagent solution*		*Boiled with pyridine, cooled, and reagent solution added*		*Pyridine and reagent solution added to sample without heating*	
	Immediate	*After 5 min.*	*immediate*	*After 5 min.*	*immediate*	*After 5 min.*
Polyvinyl chloride	Red-brown	Blood-red, Brown-red	Blood-red, Brown-red	Red-brown, black precipitate	Red-brown	Black-brown
Chlorinated PVC	Blood-red, Brown-red	Brown-red	Brown-red	Red-brown, black precipitate	Red-brown	Red-brown
Chlorinated rubber	Dark red-brown	Dark red-brown	Black-brown	Black-brown precipitate	Olive-brown	Olive-brown
Polychloro-prene	White-cloudy	White-cloudy	Colourless	Colourless	White-cloudy	White-cloudy
Polyvinyl-idene chloride	Black-brown	Black-brown precipitate	Black-brown precipitate	Black-brown precipitate	Black-brown	Black-brown
PVC molding compound	Yellow	Black-brown precipitate	White-cloudy	White precipitate	Colourless	Colourless

Table 4.12: *Colour reaction with morpholine*

Polymer	*Colour Change*
Polyvinylidene chloride (PVDC)	Gradually darkens, after several hours becomes turbid and almost black
Fluorinated polyvinyl chloride	Red — brown
Poly vinyl chloride	No colour reaction

Table 4.13: *Chlorine content in different p olymers*

Polymer	*Chlorine content (in %)*
Polyvinyl chloride	55.56
Chlorinated polyvinyl chloride	64.68
Polyvinylidene chloride	73.00
Copolymers of vinyl chloride and vinyl acetate	Always less than PVC

12. Test for PTFE, PCTFE, PVF and PVDF

(i) *Reaction on heating and burning:* PTFE does not melt, it vaporizes slowly at red heat, when heated in a micro test tube and gives off waxy smoke with stinging odour of hydrogen fluoride. PCTFE melts and decomposes at high temperature and gives of a stinging smell of hydrogen fluoride. Neither product burns or chars. The reaction of the vapours with litmus paper is strongly acidic. There is a sharp and choking odour of hydrofluoric acid for PTFE, PVF and PVDF and an additional odour of hydrochloric acid for PCTFE.

(ii) *Solubility:* There is no known solvents for PTFE, dissolves in o-dichloro benzene above 120°C, while aromatic esters and chlorinated hydrocarbons at high temperatures only cause swelling. PVF dissolves in ketones, tetra methylene sulphone and dimethyl formamide above 100°C.

Group 4: Plastics Containing Sulphur

13. Polysulphone

Polysulphones are difficult to ignite, burning continuously after ignition, with yellow sooty flame and chars. On pyrolysis it melts, bubbles, and finally forms a brownish melt. The vapours are invisible and strongly acidic in nature. It is soluble in benzene and methylene chloride and swells in acetone and ethyl acetate.

Group 5: Plastics not Containing Nitrogen, Sulphur and Halogens

14. Test for Polyolefins

The density of the olefins can be found by the technique described below:

A small piece (about 0.01 g) of the polymer is placed in a 250 ml measuring cylinder containing 100 ml of water. Methanol is added from a burette in 0.5 ml portions with vigorous stirring after each addition, until the polymer remains suspended at a constant level in the liquid (at which the density of the liquid is equal to that of the polymer). From the volume of methanol added and volume of the resultant solution, the percentage of methanol in the mixture is calculated. The corresponding density is then ascertained, from the tables given in most of handbooks of physical chemistry. (For guidance, solutions containing approximately 31, 57 and 62 volume/ volume of methanol in water have density values 0.96, 0.92 and 0.91 respectively).

A piece of dry sample is pyrolysed in a tube closed with a filter paper which is drenched with a solution of 0.5 g of yellow mercuric (II) oxide in sulphuric acid (1.5 ml concentrated sulphuric acid in 8 ml of water):

(i) If the vapour immediately gives a golden spot, it indicates the presence of polyisobutylene.

(ii) If the golden spot appears after few minutes it indicates the presence of polypropylene (PP).

(iii) Polyethylene does not react with this mixture. Different forms of polyethylenes like LDPE, HDPE, HMHDPE, etc., can be distinguished by their melting temperature.

15. Test for Styrenic (for PS, SAN, ABS, SBR) Polymers

Heat 0.5 g of plastics with concentrated nitric acid in a test tube by covering it's mouth by a filter paper soaked in a solution of 2,6-dibromo quinone in diethyl ether. Then treat the paper with ammonium hydroxide solution. Development of blue colour confirms the presence of styrene.

16. Test for Polycarbonate (PC)

(i) About 1 g of sample is pyrolysed in an ignition tube, which is plunged with cotton. The cotton is removed and immersed in 1% methanolic solution of p-dimethyl amino benzaldehyde and then one drop of 5N hydrochloric acid is added. A dark blue colour appears in the case of polycarbonates. A red colour that does not change to blue is observed in the case of polyamides.

(ii) The cotton plug in which the pyrolysed vapours of the sample absorbed is treated with dilute (1:1) hydrochloric acid, an intense red colour which is unaffected by addition of methanol indicates the presence of polycarbonate.

17. Test for PMMA

About 1 g of polymer is heated in an ignition tube and the pyrolysate is collected in a test tube wrapped with a filter paper. To the distillate, 1 ml of concentrated nitric acid is added and heated just to boil. After cooling, 5 ml of water is added and then sodium nitrite (0.1 g) is added. A bluish green colour confirms the presence of polymethyl methacrylate.

18. Test for Polyacetals (Polyoxy methylene)

Polyacetals produce formaldehyde on heating. A small amount of sample is heated with 2 ml concentrated sulphuric acid and a few crystals of chromotropic acid for about 10 min at 70°C. A strong violet colour confirms polyacetal.

19. Test for PET and PBT

PET and PBT are soluble in nitrobenzene. A small sample is pyrolysed in a glass tube covered with filter paper, which is drenched with a saturated solution of o-nitrobenzaldehyde in dilute sodium hydroxide. A blue- green (indigo) colour, which is stable against dilute hydrochloric acid, indicates the presence of terephthalic acid.

The absolute distinction between PET and PBT is difficult using simple methods. These can be differentiated based on their melting points. PET melts at 250-260°C and PBT at about 220°C. However, additives may cause deviations from these melting points. PET and PBT can be identified by the formation of white sublimate when these polymers are heated in a combustion tube.

20. Test for Unsaturated Polyesters

Unsaturated polyesters are produced in the form of dissolved resins in polymerizable monomers (usually styrene). They are also known in the form of moulding resins or as hardened products. One should distinguished between unsaturated and the saturated aliphatic and aromatic polyesters. Among the latter are PET and PBT. The acidic components of most unsaturated polyesters are maleic, phthalic, sebacic, succinic, fumaric and adipic acids. All unsaturated polyesters can be directly identified.

(i) *Test for Phthalic Acid*

(a) *Detection as phthalic anhydride:* 1 g of polymer sample is heated in a test tube. Phthalic anhydride deposits on the test tube walls in the form of needles of melting point 131°C. This confirms the presence of phthalic acid.

(b) *Detection as thymolphthalene:* 0.1 g of the sample is heated for 10 minutes in an oil bath at 120-130 °C with about 0.3 g of thymol and 1 drop of concentrated sulphuric acid. After cooling, the mixture is dissolved in 50% ethanol and made alkaline with 2N sodium hydroxide. A red colour indicates the presence of phthalic acid

(c) *Detection as phenolphthalene:* 0.01 g of the sample is carefully melted in a test tube with 0.2 g of crystalline phenol and 1 drop of concentrated sulphuric acid. After cooling, the melt is dissolved in 10 -20 ml of water and the solution is made alkaline with 5% sodium hydroxide. A red colour indicates the presence of phthalic acid.

(ii) *Tests for Succinic Acid*

A small amount of the sample is heated in a test tube to 190°C with 1 g of hydroquinone and 2 ml of concentrated sulphuric acid. After cooling, the mixture is diluted with 25 ml of water and shaken with 50-65 ml of benzene. A red colour in benzene layer indicates the presence of succinic acid. Then the benzene layer is washed with water shaken with 0.1 N aqueous sodium hydroxide. A blue colour should develop, when the test is positive.

(iii) *Test for Maleic Acid*

A few mg of the sample is dissolved in 2ml hot acetic anhydride. After cooling, 3 drops of 50% sulphuric acid are added and the developing colour is observed immediately. After 10 minutes in water bath, a wine red colour changing to olive brown indicates the presence of maleic acid.

(iv) *Test for Adipic acid with Resorcinol*

The sample is melted in a test tube with an equal amount of resorcinol and two drops of concentrated sulphuric acid. The mixture is made alkaline with sodium hydroxide. A dark purple colour develops in the presence of adipic acid. Whereas orange colour with greenish fluorescence develops in the presence of sebacic acid.

(v) *Tests for Fumaric Acid*

(a) The sample is treated with several ml of mixture composed of 4 ml of 10% aqueous copper sulphate, 1 ml of pyridine and 5 ml of water. Emerald green (greenish blue) crystal precipitate indicates the presence of fumaric acid.

(b) 0.1 g of the sample is treated with 3 ml of pyridine and 1.5 ml of acetic anhydride. A dark violet colour develops in the presence of fumaric acid at room temperature, which turns to blue on heating. A pink violet colour develops in the presence of maleic acid in the cold, which changes to brown on heating.

21. Test for Phenol-Formaldehyde (PF)

(i) *Test for Phenol*

(a) *Phenolphthalene Test:* About 1 g of the sample is heated in a porcelain evaporating dish with an equal quantity of phthalic anhydride and 3 drops of concentrated sulphuric acid is added until a dark melt is formed. On cooling, the melt is diluted with water and made alkaline with 10% sodium hydroxide solution. A characteristic red colour indicates the presence of phenol-formaldehyde. If tarry materials mask the colour, the solution is diluted with water and 1 N sulphuric acid is added drop wise. At the neutralization point a distinct change in the colour occurs.

(b) *Millon's test:* The Millon's reagent is prepared by dissolving 10 g of mercury in 10 ml of nitric acid with gentle heating followed by dilution with 15 ml of distilled water. 0.1 g of the sample is heated to boiling for 2 min with 1 ml of clear Millon's reagent. A red colour indicates the presence of phenol in the sample.

(c) *Gibb's indophenol test:* About 0.5 g of the sample is heated in a test tube. The mouth of the test tube is covered with a filter paper impregnated with 2,6-dichloro quinone-4-chloramide (filter papers are prepared by immersing the papers in benzene solution of 2,6-dichloro-quinone-4-chloramide and drying). The pyrolysis is carried out for not more than one minute. After removal from the test tube, the filter paper is placed in an ammonia vapour or is treated with 1-2 drops of dilute ammonia solution. A blue colour indicates the presence of phenol.

(ii) *Test for Formaldehyde*

A small amount of sample is boiled in water in presence of sulphuric acid and filtered. A few drops of 5% aqueous chromotropic acid solution (1,8- dihydroxyl naphthalene-3, 6-disulphonic acid) and excess of concentrated sulphuric acid are added to the filtrate and heated at 100°C for few minutes. In the presence of formaldehyde, the solution turns violet/dark violet. Polyvinyl formal, polyoxy methylene, PF, UF and MF resins contain formaldehyde.

22. Test for Epoxy Resin

(i) **Foucry Test:** About 0.25 g of sample is dissolved in 98% sulphuric acid by slight warming. After cooling, 1 ml of 63% nitric acid is added. After shaking, the mixture is poured into 10 ml of 5% aqueous sodium hydroxide. A bright red or orange red colour indicates the presence of epoxy resin.

(ii) 10 ml of 98% sulphuric acid is mixed with 50 ml of distilled water and in the still warm solution, 2.5 g of mercuric oxide is dissolved. 5 ml of this reagent is added into 1 ml of sulphuric acid-epoxy resin solution, the mixture is shaken and allowed to stand for 30 min. An orange red precipitate confirms the presence of epoxy resin.

(iii) 1 g of the sample is dissolved in 100 ml of concentrated sulphuric acid.

(a) To 1 ml of this solution, 3 drops of 40% formaldehyde solution is added. A red brown colour, which changes to green on dilution with water, confirms the presence of epoxy resin.

(b) 5 ml of 10% ferric sulphate solution are added to 1 ml of the above solution. A dark violet colour indicates the presence of epoxy resins.

23. Test for Alkyd Resins

Oil modified polyesters are called as alkyd resins. Since in most of the alkyd resins, phthalic anhydride is used as the major polybasic constituent, the presence of phthalic anhydride is identified by the following test.

Test for phthalate: 0.1g of the sample is carefully melted in a test tube with 0.3 g of crystalline phenol and one drop of concentrated sulphuric acid. After cooling, the melt is dissolved in 10-20ml water and the solution is made alkaline with 5% sodium hydroxide. A red colour indicates the phthalate, which confirms the presence of alkyd resins.

24. Test for Polyphenylene Oxide (PPO)

Poly phenylene oxide is difficult to ignite, after ignition burns continuously with bright, sooty yellow flame. On pyrolysis it melts, decomposes and becomes black. It emits brown vapours, which is alkaline in nature having slight phenolic odour. Polyphenylene oxide can also be identified by its higher melting point and density.

25. Test for Polyvinyl Alcohol (PVA)

(i) *Reaction with Iodine Reagent*

An iodine reagent is prepared by dissolving 0.1 g iodine and 1 g potassium iodide in a mixture of 10 ml ethyl alcohol and making up to 100 ml with 2N hydrochloric acid. 0.5% aqueous solution of PVA is prepared. To 1 ml of this solution, 3 drops of the iodine reagent is added. A blue colour develops, if the colouration is not observed, add 1 drop of saturated aqueous borax solution to identify the presence of polyvinyl alcohol. Polyvinyl alcohol containing more than 10% acetate groups gives a red coloration in this test.

(ii) *Borax Test*

A fairly concentrated solution of polyvinyl alcohol is prepared. One drop of this is transferred on to a spot plate and one drop of saturated borax solution is added. A gel forms in the presence of polyvinyl alcohol.

26. Test for Polyvinyl Acetate (PVAc)

(i) *Reaction with Iodine Reagent*

An iodine reagent is prepared as explained above. The sample is covered with 0.5 ml of iodine reagent. The solid becomes stained deep red in the presence of poly vinyl acetate. The effect is enhanced if 10 ml of water is added.

(ii) *Detection of Acetate*

The sample is heated with few ml of dilute hydrochloric acid and then neutralized with ammonia until slightly alkaline. 1 ml of this solution is treated with 1 to 2 drops of 0.1 N iodine solution in 0.1 N aqueous potassium iodide. A dark blue colour develops in the presence of acetates. Whereas, a brown colour develops in the presence of propionates.

(iii) *Liebermann-Storch-Morawski Reaction*

A few milligrams of the sample is dissolved in 2 ml acetic anhydride (hot). After cooling, 3 drops of 50% sulphuric acid are added. The developing colour of the solution while adding sulphuric acid, then again after 10 min and finally after heating once more to about 100°C, are observed. Change of colour from blue to red indicates the presence of polyvinyl acetate.

27. Test for Polyvinyl Acetals

A small amount of the sample is contacted with 1-2 drops of reagent solution and allowed to react for 1 min and rinsed. The developing colour is observed.

Blue to dark violet	–	Poly vinyl formal (PVF)
Blue to dark green	–	Poly vinyl acetal (PVAC)
Blue to light green,	–	Poly vinyl butyral (PVB)
Blue to dark red	–	Poly vinyl acetate (PVAc)

Reagent solution: 10 ml 50% acetic acid + 7 ml iodine in potassium iodide solution (1 g potassium iodide + 0.9 g iodine + 40 ml water + 2 ml glycerin).

28. Test for Polyvinyl Carbazole

A small amount of the sample is treated with concentrated sulphuric acid and one drop of glycerol or formaline. A sea green colour develops in the presence of polyvinyl carbazole.

29. Test for Polyvinyl Ethers

A small amount of the sample is treated with phenyl-lithium in tetrahydrofuran solvent. A red colour develops initially which subsequently turns blue, then purple and finally after some time a black gel is formed indicating the presence of poly vinyl ethers.

30. Polyesters (Thermosetting Plastics)

Polyesters burn with a yellow flame with blue edges and produce black smoke with soot. They continue to burn without dripping after removal of the ignition source and give off a sour cinnamon odour. The specific gravity ranges from 1.30 to 1.50.

31. Silicones (Thermosetting Plastics)

Silicones burn with a bright yellow flame without odour and continue to burn after removal from the flame source. The specific gravity ranges from 1.05 to 2.82. The clear identification is made by an IR spectrometer. To test for silicon, mix approximately 30 mg of the sample with 100 mg sodium carbonate and 10 mg sodium peroxide. Heat this in a platinum or nickel crucible over a flame. Dissolve the melt in a few drops of water, boil it, and then add dilute nitric acid until the solution is neutral or slightly acidic. This is followed by the addition of few drops of ammonium molybdate solution and heating to near boil. Cool the sample, add 1 drop of benzadine solution (50 mg benzadine dissolved in 10 ml 50% acetic acid; add water to make 100 ml) and then add 1 drop of saturated aqueous sodium acetate solution. A blue colour formation indicates the presence of silicon.

32. Diallyl Phthalate (DAP)

DAP compounds specific gravity ranges from 1.80 to 1.85 depending upon, the type of filler. It burns with a yellow flame, produces black smoke, and gives off the faint odour of phenol. It is self-extinguishing in nature and is difficult to ignite.

33. Shellac

About 0.05 g of polymer is dissolved in 1 ml of ethanol with warming. To the cooled solution add 1 ml of water to give a white emulsion. It is treated with 2 drops of aqueous NaOH. Appearance of violet red colour confirms the presence of shellac. The colour change is reversed on acidification with 5N HCl.

34. High Temperature-Resistant (HT) Thermoplastics

The term high temperature-resistant (HT) thermoplastics are used for polymers, which in the absence of filler have a continuous-use temperature above approximately 200°C. Among the most important HT thermoplastics are: polyphenylene sulphides (PPS), polysulphones (PSU), polyether sulphones (PES), polyether imdies (PEI), PEEK and poly arylates (PAR). A first approach to differentiate the most important high temperature thermoplastics is to examine their behavior in various solvents. Table 4.14 shows that PEEK and PES are completely insoluble in the most common solvents, while PAR, PEI and PES dissolve easily in chlorinated solvents. Polyamide PA 6-3-T (polytrimethyl hexamethylene terephthalate) does not dissolve in chlorinated solvents, but is soluble in dimethyl formamide. Among high temperature (HT) polymers, which are insoluble at room temperature, one can recognize PPS by the fact that over approximately 210 °C it dissolves in chloro naphthalene and methoxy naphthalene. Further information can be obtained from the pH value of the pyrolysis vapours (see Table 4.5). With PAR, one also notices a smell of phenol that is an indication of the presence of phenolic components in the polymer chain. Pyrolysis of PES results in a stinging odour of SO_2, pyrolysis of PPS results in an odour of H_2S; PEI and PA 6-3-T on heating smell like burnt horn. The behavior of HT thermoplastics when they burn can also be used for their identification (see Table 4.15).

Additional information can be obtained from certain special tests. Thus the Gibb's Indophenol tests is positive with PEEK, PAR and PEI (see Test No. 21 (i) (c)). Using the color reaction with p-dimethyl amino benzaldeyhyde one can differentiate HT-polyamide such as PA 6-3-T, from polymers which develop phenolic decomposition products during pyrolysis. While with polyamides the red coloration obtained after addition of conc. HCl, remains, with PC it turns blue.

Table 4.14: *Dissolution temperature of high temperature resistant thermoplastics (°C)*

Solvents / Polymers	Chloro form	Tetra hydro furan	p-Xylene	Dichloro methane	Dimethyl form amide	m-Cresol	Ethyl acetate	Trichloro benzene	Methyl ethyl ketone	Toluene
PEEK	i	i	i	i	i	sw	i	i	i	i
PEEK (amor phous)	25	25	i	20	138	40	i	30	i	i
PEI	25	sw	i	20	45	58	i	110	i	i
PA 6-3-T	i	i	i	i	30	60	i	20	i	i
PAR	25	25	i	20	111	40	i	30	i	i
PESU	25	i	i	20	20	40	i	i	i	i
PPS	i	i	i	i	i	i	i	i	i	i

[Note: i - insoluble; sw - swellable]

The burning behaviour of HT-thermoplastics is given in Table 4.15.

***Table 4.15** : Burning behavior of HT-thermoplastics*

Polymer	*Flammability*	*Flame*	*Odor of Smoke*	*Indication*
PEEK	Extinguishes soon after removal from flame	Slightly yellowish, slightly smoky	Slight of phenol	Phenol
PEEK (amorphous)	Extinguishes immediately after removal from flame	Yellow, foams	Slight of phenol	Phenol
PEI	Extinguishes immediately after removal from flame	Slightly yellowish, foams	Burnt horn	Nitrogen (amides, imides)
PA 6-3-T	Extinguishes immediately after removal from flame	Blue flame border, very sooty	Burnt horn	Nitrogen (amides)
PAR	Extinguishes immediately after removal from flame	Yellow, foams	Phenol	Phenol
PESU	Extinguishes immediately after removal from flame	Yellow, foams	Stinging of sulfur dioxide	Sulfur
PPS	Extinguishes immediately after removal from flame	Very sooty	Strong of hydrogen sulfide	Sulfur

MISCELLANEOUS PROPERTIES OF PLASTICS

Typical properties such as physical, optical, mechanical, thermal electrical properties and applications of some plastics are given in Tables 4.16 to 4.19. The explanation for some of the common terminologies frequently used in polymer field are highlighted below.

1. Solubility Parameter

The term solubility parameter ($\Delta E/V$) indicates the amount of energy required to vaporize a unit of liquid. In the process of vaporization, energy is spent to overcome the cohesive force holding the molecules together and to let the molecules completely separate out into the gaseous state. The term ($\Delta E/V$), therefore, is a measure of the cohesive force that holds the molecules in the liquid together and has been designated as the "Cohesive Energy Density" (CED). Solubility parameter is defined as the square root of the CED:

$$\delta = \left(\frac{\Delta E}{V}\right)^{\frac{1}{2}} \quad (\text{unit : cal}^{1/2}/\text{cm}^{3/2})$$

2. Optical Properties

(a) Refractive index (ASTM D542): Refractive index (RI) is a fundamental property of transparent materials. RI values are very important for design engineers involved in designing lenses for cameras, microscopes and other optical instruments. RI is defined as the ratio of velocity of light in vacuum (in air) to its velocity in a transparent medium.

$$\text{Refractive index} = \sin I/\sin R$$

where, 'I' is the angle of incidence and 'R' is the angle of refraction.

(b) Haze (ASTM D1003): The cloudy or turbid aspect of appearance of an otherwise transparent specimen caused by light scattering from within the specimen or from its surface.

Haze of a specimen is defined as the percentage of transmitted light that, is passing through the specimen, deviates more than 2.5° from the incident beam by forward scattering.

(c) Luminous transmittance (ASTM D791): Luminous transmittance is defined as the ratio of transmitted light to the incident light. Both properties vary with the geometry of the optical measuring system, which must be carefully specified. The three optical properties of interest are:

$$\text{Total transmittance } (T_t) = T_1 - T_2 \;/\; T_1$$

$$\text{Diffuse transmittance } (T_d) = (T_4 - T_3) \times (T_2/T_1) \;/\; T_1$$

$$\text{Haze } (\%) = T_d \;/\; T_t \times 100$$

where, T_1, T_2, T_3, and T_4 represents the incident light, total light transmitted by the specimen, the light scattered by the instrument and the scattered light by the specimen, respectively.

(d) Gloss: Gloss is a term employed to describe the surface character of a material which is responsible for luster or shine, i.e., surface reflection of a polymeric material.

3. Thermal Properties

(a) Heat capacity (C): Heat capacity is the ratio of differential heat energy (dQ) to differential temperature (dT) for a material.

$$C = dQ \ / \ dT$$

(b) Specific heat, (c): It is defined as the heat capacity (C) per unit mass or per unit volume. Typical specific heat values for solids, including plastics, range from 0.4 J/gK to 4 J/gK.

Specific heat change with the temperature. In crystalline polymers it is maximum at the melting point.

(c) Thermal conductivity (ASTM DC177): Thermal conductivity is defined as the rate at which heat is transmitted by conduction through a unit cross sectional area of material when a temperature gradient exists perpendicular to the area.

$$K = \frac{Q \times 1}{[A(T_1 - T_2)]}$$

where, Q is heat supplied; K is thermal conductivity; t is thickness of specimen; A is area of specimen; T_1 is temp. of hot surface; and T_2 is temp. of cold surface. Low thermal conductivity is one of the most important features of plastics. On one side it has made processability more difficult, but at the same time this property has made it possible to use plastics as excellent insulators.

(d) Coefficient of expansion: The fractional change in dimension (some times volume) specified, of a material for a unit change in temperature. It can be expressed in mm/mm°C (ASTMD 696).

4. Electrical Properties

(a) Volume resistivity (ρ_v): The volume resistivity of a material is the ratio of the potential gradient parallel to the current in the material to the current density (ohms.cm)

$$\rho_v = R_v A/t$$

where R_v = volume resistance; A = area of electrode in sq.cm; t = thickness of the specimen, cm.

(b) Volume Resistivity (ρ_s): The surface resistivity of a material is the ratio of material to the current per unit width of the surface (ohms).

$$\rho_s = R_s P/g$$

where R_s = surface resistance; P = Perimeter of gaurd electrode in cm; g = gap between the electrodes, cm.

(c) Electric Breakdown: Failure in the function as an insulating material, especially in preventing passage of an electric discharge, is termed as electrical breakdown. It is manifested by an unstable increase in current flow with time, or by physical disruption of the material.

(d) Dielectric Strength: Dielectric strength is the ratio of dielectric breakdown voltage to the thickness of an insulating material (volts/mil). This is also known as electric strength and breakdown strength.

(e) Dielectric Constant: The dielectric constant defines the ability of a material to store electrostatic charge. It is the ratio between the condenser capacity when filled with a dielectric material and the same in vacuum. The dielectric constant is also termed as specific inductive capacity and permittivity.

(f) Dissipation Factor (Loss tangent): The measure of power loss dissipated as heat due to the build-up and collapse of the electric field within the insulation in an a.c. circuit, especially at high frequencies, is termed as dissipation factor.

Table 4.16 : *Density, solubility parameter and optical properties of some plastics.*

Polymers	ρ_c *(g/cc)*	ρ_a *(g/cc)*	ρ_c/ρ_a	*Solubility parameter (δ) (cal.cm^{-3})$^{1/2}$*	*Light transmittance (%)*	*Haze (%)*	*Refractive index*
Polyolefins							
Poly propylene	0.95	0.85	1.118	8.22–9.2	55–90	1.35	1.490
Polyethylene	1.00	0.85	1.176	7.72–8.35	10–80	4–50	1.516
Styrenes							
Poly styrene (PS)	1.13	1.05	1.08	8.5–9.3	87–92	<0.1–3	1.57–1.6
Poly-α-chlorstyrene	—	<1.25	—	8.9	—	—	1.610
Styrene acrylonitrile (SAN)	—	—	—	—	78–88	0.4	1.56–1.57
Poly-acrylonitrile (PAN)	1.18	1.17	1.01	12.5–15.4	90	—	1.514
Polyvinyl alcohol (PVA)	1.35	1.26	1.07	12.6–14.2	—	—	1.500
Polyphenylene oxide (PPO)	1.15	1.00	1.15	—	—	—	—
Polyvinyl acetate (PVAc)	>1.19	1.19	—	8.85–11.05	—	—	1.467
Poly amides							
Nylon 6	1.23	1.08	1.14	—	—	—	1.53
Nylon 6,6	1.24	1.07	1.16	13.7	—	—	1.53
Nylon 6,10	1.19	1.04	1.14	—	—	—	1.53

Acrylates							
Polymethacrylate (PMA)	—	1.22	—	9.7–10.4	—	—	1.479
Polymethyl methacrylate (PMMA)	1.23	1.17	1.05	9.1–12.8	—	—	1.050
Halogenated Polymers							
Polyvinyl chloride (PVC)	1.52	1.39	1.10	9.4–10.8	76-82	8–18	1.539
Polyvinylidene chloride (PVDC)	1.95	1.66	1.17	9.9–12.2	—	—	1.6–1.63
Polyvinylidene fluoride (PVDF)	2.00	1.74	1.15	7.2–7.9	—	—	1.420
Polychlorotrifluoroethylene (PCTFE)	2.19	1.92	1.14	—	—	3–5	1.425
Polytetrafluoroethylene (PTFE)	2.35	2.00	1.17	6.2	—	—	1.35–1.38
Engineering Polymers							
Polyethylene terephthalate (PET)	1.46	1.33	1.10	9.7–10.7	—	—	1.51–1.65
Polycarbonate (PC)	1.35	1.12	1.21	—	85–91	0.5–2	1.58–1.59
ABS	—	—	—	—	28	100	1.520

Table 4.17: *Mechanical properties of some plastic materials*

Plastic materials	*Impact strength (kg. cm/cm) Notched*	*Tensile strength* $(kg/cm^2) \times 10^2$	*Tensile modulus* $(kg/cm^2) \times 10^3$	*Flexural strength* $(kg/cm^2) \times 10^2$	*Compressive strength* $(kg/cm^2) \times 10^2$	*Water absorption at 23°C for 24 hrs (%)*
Ethyl vinyl acetate	No Break	0.63–2.86	0.21–1.06	—	—	<0.01
Low density polyethylene	No Break	0.70–1.75	1.40–3.16	—	—	<0.05
High density polyethylene	2.72–125.12	1.90–3.09	4.23–12.68	0.070–2.81	0.56–2.54	<0.01
High Mol Wt High-density polyethylene	108.8	2.03–4.22	112.2	2.46	1.69	<0.01
Polypropylene	2.72–81.6	3.62	14.08	3.16–5.63	4.22–7.04	<0.01
Polystyrene	1.36–2.18	3.30–5.76	31.65–33.42	6.33–10.56	8.10–11.27	0.03–0.05
High impact polystyrene	3.18–19.04	1.63–4.21	16.86–26.00	3.87–8.80	5.63–11.27	0.05–0.10
Polyvinyl chloride (Flexible)	Varies	0.56–2.67	0.049–0.126	—	0.63–1.20	0.15–0.75
Polyvinyl chloride (Rigid)	2.18–119.68	4.18–5.30	—	5.63–10.57	7.04–7.75	0.07–0.40
Polyvinylidene chloride	8.16–38.08	3.6	5.27	10.00–11.97	9.15–15.49	0.07–0.11
ABS	7.07–54.40	3.37–4.39	16.83–23.15	3.52–9.51	3.52–7.75	0.2–0.5
Acetal	5.98–7.62	6.62	28.85	9.15–9.86	12.68	0.25
Acrylic	1.63–12.51	4.90–7.76	22.86–33.05	9.86–11.97	9.86–11.97	0.2–0.3
Nylon 6	4.90–10.88	8.16	30.6	6.34–11.69	2.82–2.75	1.5
Nylon 6,6	4.90–10.88	9.18	34.68	10.28	3.52–9.15	1.3
Polycarbonate	10.88–16.32	6.32	24.27	7.75–9.15	8.80	0.15

Polyethylene terephthalate	9.25	5.10	17.58	—	—	<0.1
Polyphenylene oxide	8.16–10.34	4.90	25.30	10.56	10.56	0.06
Polyurethane (thermoplastic)	No Break	1.05-7.75	0.023–0.10	0.49–0.70	>14.08	0.06–0.80
Styrene acrylonitrile	1.63–2.72	5.81-7.65	34.88	11.97	10.56–12.32	0.23–0.28
Polytetrafluoroethylene	No Break	2.11	4.21	—	2.82–8.45	0.01
Polyacrylate	22.85	6.73	20.4	1.75	—	0.05–0.25
Polyaryl sulphone	5.44–10.88	8.44	27.08	12.11	12.61	1.1
Polyetherimide	5.04–11.42	10.71	30.25	—	—	0.13
Polyether ether ketone (PEEK)	8.43–9.76	9.38	36.72	—	—	0.5
Polyether sulphone (PES)	8.70	8.75	24.89	13.10	—	0.43
Polyphenylene sulphide	1.63–4.35	12.34	33.66	—	—	0.01–0.02
Melamine-formaldehyde	1.31–7.62	5.61	94.86	7.75–9.86	28.17–31.69	0.1–0.4
Phenol-formaldehyde	1.1–24.48	6.02	105.06	8.45–10.56	7.04–21.13	0.4–0.5
Polyurethane (thermoset)	136–No Break	9.18	4.18	—	—	0.2–1.5

Table 4.18: *Thermal and Electrical Properties of some Plastic Materials*

Plastic materials	*Coefficient of thermal expansion (mm/mm/°C × 10^{-5})*	*Thermal conductivity (cal/(S) cm^2/°C/cm × 10^{-4})*	*Dielectric strength (KV/mm)*	*Dielectric constant at 60 Hz*	*Volume resistivity, Ω-cm*
Ethyl vinyl acetate	—	8	21	3.16	1.5×10^{8}
Low density polyethylene	18–39.6	8	24	2.28	10^{15}-10^{18}
High density polyethylene	10.8–19.8	1.9-3.3	24	2.3	6x10^{15}-10^{18}
High Mol Wt high density polyethylene	13–20	8	20	2.3-2.6	>10^{16}
Polypropylene	9	2.8-4	28	2.1-2.7	10^{15}-10^{17}
Polystyrene	3.4–10.1	1-3	20	2.5-2.65	10^{17}-10^{21}
High impact polystyrene	5–8.5	8	20	2.5-3.5	10^{10}-10^{17}
Poly vinyl chloride (Flexible)	1.26–4.5	3-4	16	5.9	10^{11}-10^{15}
Poly vinyl chloride (Rigid)	9–18	3-5	17	3.4	10^{12}-10^{16}
Poly vinylidene chloride	7–8	3.0	-	3.08	10^{15}
ABS	9.5–11	4-9	24	2.6-3.5	10^{15}-10^{17}
Acetal	10–15	1.6	20	3.7-3.8	1-10^{14}
Acrylic	5–9	1.4	20	3.3-3.9	>10^{14}
Nylon 6	8.1	5.9	20	6.1	10^{14}-10^{15}
Nylon 6,6	8.1	5.8	24	3.6-4.0	10^{14}-10^{15}
Polycarbonate	12.2	4.6	15	2.97	6.1×10^{15}

Polyethylene terephthalate	6.0	4.2	26	3.65	3×10^{16}
Polyphenylene oxide	7.2	—	25	2.58	10^{17}
Polyurethane (thermoplastic)	10.0	7.4	40	6-8	2×10^{11}
Styrene acrylonitrile	67.0	3.0	20	2.8-3.0	10^{15}
Polytetra fluoroethylene	9.9	6.0	25	2.1	$>10^{18}$
Polyacrylate	5-6.2	7.13	16	2.62	—
Polyaryl sulphone	4.9	4.55	15	3.94	3.2×10^{16}
Polyetherimide	5.6	—	33	3.1	—
Polyether ether ketone	4.7	—	19	3.2	4.9×10^{16}
Polyether sulphone	5.5	3.2-4.4	16	3.5	10^{17}-10^{18}
Polyphenylene sulphide	2.9	6.84	24	3.1	10^{16}
Melamine-formaldehyde	4.0	7.0-8.0	12	7.9-9.5	10^{12}-2×10^{12}
Phenol-formaldehyde	2.5-6.0	3.0-6.0	9	5.5	10^{9}-10^{13}
Polyurethane (thermoset)	32-65	—	9	—	2×10^{11}-10^{14}
Urea-formaldehyde	2.0-5.0	7.0-9.0	14	7-9.5	10^{12}-10^{13}
Epoxy	4.5-6.5	4.0-5.0	20	3.5-5.0	$>10^{14}$

Table 4.19: *Properties and applications of some important plastics*

Plastic material	*Properties*	*Applications*
Polyethylene (PE)	Soft, flexible, translucent	Packaging, extrusion, coating
LDPE, HDPE	Soft, flexible, transparent	Bottles, packaging, toys, films
LLDPE, HMHDPE	Stiff, rigid, opaque	Blown films, packaging
UHMWPE	Stiff, rigid, opaque	Composites, automobile fuel tanks, aerospace
VLDPE	Soft, flexi ble, transparent	Vehicle parts, thin films, shrink wrappings
Polypropylene (PP)	Flexible, translucent, tough, good chemical resistance	Extensively used in packaging and automobile parts, blown films, toys, disposable syringe, and rope.
Reinforced PP	Good tensile strength, good abrasion resistance, high HDT	Composites in automobile parts, radiator grills, housings.
Polystyrene (PS)		Cups, decorative articles, encapsulations, fridge lining
GPPS	Hard, brittle, transparent	
HIPS	Soft, tough, translucent	
ABS	Stiff, flexible, tough, good chemical and alkali resistance, opaque, low temperature resistance	Radio and TV cabinets, toys, housings, refrigerator lining, automobile parts, storage bins.
SAN	Good clarity, chemical resistance, brittle.	Light housing, radiator grills, and refrigerator parts.

Polyvinyl chloride (PVC)	Weatherability, inflammable, electrical insulation	Pipes, floorings, footwear, wire and cable coatings, sheets.
EVA	Rubbery, flexible, tough, good puncture resistance	Sound deadening sheets, packaging co-extrusion, bonding adhesive.
PMMA	Good gloss, weather resistance, optical clarity	Decorative lamp covers, aircraft glazes, contact lenses, windows.
Nylon 6/ Nylon 6,6	High tensile strength, stiff, flexural strength, crystalline, good oil resistance.	Major textile fiber, electrical and electronic components, casted articles, reinforcing fiber bearing, gears, machine parts
PET/PBT	Optical clarity, good strength, crystalline, good chemical and oil resistance, non-toxic	Bottles (stretch blow), containers, boil-in bag casseroles, decorative domes, sign boards
Polyacetal (POM)	Very high tensile strength and stiffness, excellent solvent, chemical & abrasion resistance,	Appliance parts, hard wares, gears, bushings, aerosol bottles, rollers, bearings
Polycarbonate (PC)	High tensile & impact strength, optical clarity, high HDT	Computer cabinets, amorphous feeding bottles, large housings
Cellulose acetate	Optical clarity, moisture resistance, puncture resistance	Tool handles, spectacle frames, photo graphic films, combs
PTFE	Self lubricating, high thermal stability, inflammable	Bearings, gears, heat resistance coatings, non-stick cook wears

(*contd.*)

Plastic material	*Properties*	*Applications*
Polyurethane (PU)	Highly elastic, good impact strength and solvent resistance	Car bumpers, RIM moldings, swim suit, coatings, foams
LCP	Lowest shrinkage, excellent chemical resistance and strength	High performance optical fibres, thixotropic liquids
PEEK	Very high HDT, good strength	Aerospace vehicle parts, tanks
PPO	Tough, good fatigue resistance	Engineering blends, machine parts
PPS	Good acid resistance, high HDT, flame retardant, good hydrolytic and chemical resistance	Electrical housings, laminates, impellers pump parts, exterior light reflectors in automobiles
PF	High heat & acid resistance, good electrical properties	Handles of cook ware adhesives, varnishes, TV and telephone housings
UF	Light colors, excellent electrical properties, acid proof	Coatings and adhesives, lacquers, crockery, electrical switches
MF	Same as UF	Dinner ware, electrical goods
Epoxies	Dimensional stability, good mechanical properties	Composites in aerospace and aircraft, laminates, adhesives, paints
USP	Most versatile and economical, good electrical & mechanical properties.	Coatings, flexible frame, shoe soles, rigid foam, composites, laminates

Table 4.20: *Primary Fillers for Plastics*

Filler	*Plastic*
Asbestos	Phenolic plastics, amino plastics, casting resins, polyesters, epoxide resins, silicones, recently also for thermoplastics such as polyethylene, polypropylene, polyvinyl chloride, and fluorocarbons
Mica	Phenolic plastics, amino plastics
Quartz powder	Phenolic plastics, amino plastics
Silica (silicon dioxide)	Phenol resins (foundry sand core), polyesters, epoxide resins. Used in small amounts in polyethylene and polyvinyl chloride to improve the flow
Glass	Phenol resins, polyesters, allyl resins, amino plastics, silicones
Chalk (calcium carbonate)	Thermoplastics such as polyvinyl chloride, polyethylene, as well as polyesters, epoxide resins, polyurethane foam
Porcelain clay (aluminum silicate)	Polyesters, epoxide resins, polyvinyl chloride
Calcium silicate	Polyesters, hardening resins
Titanium dioxide	Polyesters, epoxide resins, thermoplastics, hardening resins
Wood flour	Phenolic plastics, amino plastics
Wood pulp (paper fibers, shredded paper, paper webs)	Phenolic plastics, amino plastics, laminated plastic products, thermoplastics
Cotton (fibres, macerated cotton fabric, fabric web)	Phenolic plastics, amino plastics, laminated plastic products
Synthetic fibres	Phenolic plastics, allyl resins
Carbon black	Polyethylene (also as UV light stabilizer), butadiene-styrene copolymers, natural rubber, synthetic rubbers.

Table 4.21: *Plastic surface cleaning*

Plastic	*Method*
Acrylics	Wipe with methanol
Aminoplasts	Scrub (detergent), scour, rinse
Cellulosics	Wipe with alcohol, scour
Epoxy	Degrease (acetone, MEK)
Flurocarbon	Sodium (solubilised), acetone wash
Nylon	Phenol (12% aqueous), 1:1 resorcinol-ethanol
Polyamide	Solvent-clean (acetone, MEK), Scrub
Polyolefine	Oxidise (chromate or radiation)
Polystyrene	Wipe with alcohol
Polyurethane	Wipe with acetone or MEK
Rubber (NBR, SBR, neoprene)	Scrape, scrub with toluene (immerse in acid)
Reinforced plastics	Wipe with MEK

(MEK – Methyl ethyl ketone)

Table 4.22: *Recommended adhesives for bonding plastics*

Plastic	*Adhesives*
ABS	Epoxy, urethane, acrylic, nitrile-phenolic, cyanoacrylate
Acetal	Epoxy, phenolic, polyester, EVA, cyano acrylate
Cellulose acetate	Urethane, resorcinol-formaldehyde, nitrile-phenolic, rubber based
Elastomers	Pressure-sensitive based on similar elastomer, urethane
Epoxy	Epoxy (with primers), nitrile-phenolic, acrylic, polyester, resorcinol-formaldehyde
Flurocarbon	Urethane
Nylon	Phenolics, epoxy, polyamide hot melt
Phenolic	Epoxy, polyvinyl acetate, urea-formaldehyde, acrylic, urethane
PMMA	Cyanoacrylate, nitrile rubber, phenolics, epoxy, urethane

(contd.)

Plastic	*Adhesives*
Polycarbonate	Epoxy, acrylic, urethane, silicone, cyanoacrylate
Polyolefines (untreated)	Rubber based, EVA, modified polyethylene
Polyester (linear)	Polyester, cyanoacrylate, nitrile rubber, urethane, acrylic
Polyester (unsaturated)	Acrylic, urethane, neoprene, nitrile-phenolic, epoxy
Polysulphone	Epoxy, vinyl-phenolic, rubber-based, urethane, polyester, cyanoacrylate
Polystyrene	Vinyl acetate-vinyl chloride emulsions, acrylic, polyamide, urethane, epoxy, polyester, hot melt, urea-formaldehyde
Polyurethane	Pressure-sensitive rubber emulsion, polyester, epoxy, phenolic, urethane, nitrile rubber
PVC (flexible)	Nitrile rubber, neoprene, urethane, epoxy
PVC (rigid)	Epoxy, urethane, acrylic, nitrile rubber, silicone, nitrile-phenolic
Silicone	Silicone
Urea–formaldehyde	Epoxy, nitrile-phenolic, phenolic, polyester, neoprene, cyanoacrylate

Table 4.23: *National and international trade names and manufacturers of some plastics*

Plastic material	*Indian*		*International*	
	Trade name	*Manufacturer*	*Trade name*	*Manufacturer*
Polyethylene (PE)	Indothene (LDPE) Othene (LDPE) Hostalen (HDPE) Escorene (LLDPE)	Indina Petro Chemical Corp. Oswal Agro Petro-Chemicals Polyolefine Industries Ltd. India Petro Chemical Corp	Amoco (LDPE) Althon (HDPE) Hostalen (HDPE) Marlex (LLDPE)	Amoco Chem. Ltd. DuPont Co. Hoechst Celanese Corporation Philips Co.
Polystyrene (PS)	Polystyron (GPPS) Styron (HIPS) Methyl styrene	Polychem. Ltd. —do— McDowell Polymers Ltd.	Styropor-BF Thermocomp Hostyrene	BASF Corporation ICI Advanced Matl. Hoechst Celanese Corporation
ABS	Polyac Acripol Absolac Anstron	Polychem. Ltd. Gujrath Chem., ABS Plastics Shansali Engg., Polymers Ltd.	Marnum Cycolac Kralastic Lustron	Dow Plastics Marbon US Rubber Monsanto
Polypropylene (PP)	Koylene Koylene-CP (PPCP)	Indian Petro Chemical Corp.Ltd., —do—	Escorene Frontline	Exxon Chem. America, Soltex Polymer Corp.

Reinforced PP	Pilene -PPR TIPCO	Polyolefins Industries Ltd., B.M. Thakkar Co.		
EVA	Pilena-EVA	Polyolefines, Industries Ltd.	Elvax Eval	DuPont Co. Hercules
PMMA	Acrypol	Gujarath State Fertilizer Corp.,	Lucite Flexi glass	DuPont Co., Rohm and Hass Co.
Nylon 6/ Nylon 6,6	Tufnyl Akulon Gujlon Sation (CGM and CGS)	Sriram Fibres Cenka Plastics, Gujarath State, Fertilizer Corp., Nymold Pvt. Ltd., Garware Polymers	Ultramid Zytel Grilon Vastmaid	BASF corporation, DuPont Co. EMSIO Ltd., Huls-America
PET	Futura and Amite	Indian Organic Chemicals Ltd. Cenka Plastics	Petra Aspect Impet	Allied Signal Co. Phillips Co. Hoechst Celanese Corp.
PBT	Arnite	Cenka Plastics	Valox Vandar	GE Plastics, Hoechst Celanese Corporation
Polyacetal			Delrin Ultraform Celon	DuPont Co. BASF corporation Hoechst Celanese Corporation

(contd.)

Plastic material	*Indian*		*International*	
	Trade name	*Manufacturer*	*Trade name*	*Manufacturer*
Polyvinyl chloride (PVC)	Indovin Reon	IPCL Reliance Ltd.	Vinylon Geon	Amoco perfrct prodn. Goodrich
Polycarbonate (PC)	-	-	Lexan, Calibre	GE Plastics Dow Plastics
Cellolulose acetate (CA)	Mycel Indac	Mysore Acetates, Eastman Anglia Plastics Ltd.	Tenite Acele Arnel	Eastman Chem., DuPont Celanese
PTFE	Hiflon	Hindusthan Fluoro Carbons Ltd.	Teflon, Fluon	DuPont Co. ICI Americas Inc.
Polyurethane (PU)	Uthane	Chemicals and Plastics (I) Ltd.	Pellethane Estane	Dow Plastics BF-Goodrich
LCP	-	-	Vectra	Hoechst Celanese, Corporation
PEEK	-	-	Victerx	ICI Advanced Matl.
PPO	-	-	Noryl	GE Plastics

PPS	-	-	Forton Ryton	Hoechst Celanese Corp. Phillips Co.
SAN	-	-	Lustran, Luran	Monsento Co., BASF Corporation
PF	Bakelite Florite	Bakelite Hylam Ltd. Indian Plastics Ltd.	Plenco Bakelite Durez	Plastics Engineering Union Carbide Hooker
UF	Presrite Florite	Bakelite Hylam Ltd. Indian Plastics Ltd.	Avisco Beetle	American Viscose American Cyanamide
MF	Florite Epoxide	Indian Plastics Ltd. SIP Resins Ltd.	Plaskon	Allied
Epoxies	Epoxite	SIP Resins Ltd.	Araldite Epin	Ciba Geigy (I) Ltd. Shell
USP	Hylak	Bakelite Hylam Ltd., Hyderabad	Gen-Glaze Selectron	General Tire PPG Industries
Diallyl Phthalate			Durez	Hooker
Alkyd, diallyl phthalate			Plaskon	Allied
Silicone			Silasic Rimplast	Dow Petrach Systems

Part B

Identification of Rubbers

The identification of unknown rubber sample may be quite complex, especially when the sample is in the compounded or unvulcanized state. The ingredient present can also give a positive response to certain tests.

Rubbers can also be identified by physical and chemical methods. It involves five simple steps. They are preliminary test, solubility test, transition temperature, elemental analysis and confirmation/chemical tests. The systematic qualitative analysis of rubbers are described below:

I. PRELIMINARY TESTS

1. Visual Appearance

This mainly deals with the observations of the sample for appearance to identify whether it is in the raw or compounded and vulcanized state. As a general rule the color and form will differentiate between these possibilities.

Raw rubbers are commonly produced as sheets or granules, which are pressed in to bales. Rubbers are commonly compounded on a two roll mill from which they are sheeted. Rubbers are soft to chew. They are embossed, it taken in the form of sheets. SBR and BR show some translucency in pure form. The elastomers acronym, starting materials, typical repeating units and their grades are given in Table 4.24.

2. Rigidity Test

Rigidity distinguishes between a raw and vulcanized state. Rubbers may be bent to a maximum extent without fracture. They can be highly extended and, if vulcanized, will have high modulus and dimensional stability.

3. Heating Test

Uncompounded rubber is infusible. Compounded rubbers cure when heated. Hence, we can clearly distinguish between compounded and uncompounded rubber by this test. All raw rubbers are lighter than water.

Table 4.24: *Abbreviations, starting materials and reactive groups of elastomers*

Chemical or technical elastomers	*Acronym*	*Starting materials*	*Typical repeating units*	*Isomers/grades*
Polybutadiene	BR	Butadiene	$-[CH_2-CH=CH-CH_2-$ $-CH_2-CH(CH=CH_2)-$	1,4-Addition (cis or trans), 1,2-Addition (isotactic, syndiotactic, or atactic)
Polychloroprene (Neoprene, Perbunan)	CR	Chloroprene	$[CH_2\text{-}C(Cl)=CH\text{-}CH_2]_n$	Structural isomers exists
Polyisoprene	PIP NR	Isoprene Natural rubber	$\text{-}CH_2\text{-}C(CH_3)=CH\text{-}CH_2\text{-}$	Cis-1,4-Polyisoprene, (Gutta percha or Balata: Trans-1,4-Polyisoprene)
Nitrile rubber	NBR	Acrylonitrile and butadiene	$-(CH_2-CH(CN))_n-(CH_2-CH=CH-CH_2)_m-$	Medium ACN High ACN Low ACN
Styrene-butadiene rubber	SBR	Styrene and butadiene	$(CH_2-CH(C_6H_5))_x(CH_2-CH=CH-CH_2)_y$	Hot SBR and Cold SBR
Butyl rubber	IIR	Isobutylene and a small amount of isoprene	$(H_2C-C(CH_3)_2-CH_2-C(CH_3)=CH-CH_2)_n$	Low unsaturation (0.6-1.2 mole %) Medium unsaturation (1.5-2 mole %) High unsaturation (above 2 mole %)

(*contd.*)

Chemical or technical elastomers	Acronym	Starting materials	Typical repeating units	Isomers/grades
Ethylene-propylene rubber	EPM	Ethylene and propylene	$-[(CH_2-CH_2)_x(CH(CH_3)-CH_2)_y]-$	
	EPDM Or EPD	Ethylene and propylene terpolymers with dienes	$-[(CH_2-CH_2)_x(CH(CH_3)-CH_2)_y(\text{diene})_z]_n-$	
Fluorine rubber	FE	Fluorine-containing olefins	Vinylidene fluoride/hexa-fluoro propene (VF_2/HFP) and VF_2/HFP + TFE	—
Chlorohydrin rubber	CHR	Epichlorohydrin-ethylene oxide copolymers	$[CH(CH_2Cl)-CH_2-O-CH_2-CH_2-O]_n$	Hydrin 100, Hydrin 200 & Hydrin 400
Propylene oxide rubber	POR	Copolymers from propylene oxide and allyl glycidyl ehter	—	—
Polyurethane rubber	PU	Macrodiol + diisocyanate + Chain extender	$-(NHCOO-R'-COONH-R)_n-$	
Polyacrylic elastomers		Acrlic monomers with bulky pendent groups	$-(CH_2-CH(C(=O)-OC_nH_{2n+1}))_m-$ If $n = 2 \rightarrow$ ethyl acrylate $n = 4 \rightarrow$ butylacrylate	

A small piece (about 0.1 g) of the material is placed on a clean spatula, previously heated to remove any traces of combustible material. It is then gently warmed, without ignition, over a small colourless Bunsen flame until it begins to fume. The sample is then removed from the flame and the odour of the fumes and whether they are acid, alkaline or neutral (damp litmus paper) is ascertained. It is important that the analyst is familiar with the odour, which the various polymers give on heating, and if necessary the odour given by the unknown sample should be compared directly with that given by authentic samples.

The sample is now moved to the hottest zone of the small Bunsen flame and the following points noted:

(i) Whether the material burns or not and, if so, how easily.

(ii) The nature and colour of the flame (sooty flame = aromatic polymer, or may be due to carbon black filler).

(iii) Whether or not the material continues to burn after removal from the flame.

(iv) The nature of the residue.

4. Bounce Test

All rubbers bounce due to resilience effect but butyl rubber does not.

5. Burning Test

Below Table 4.25 gives the typical characteristics of burning of raw rubbers.

6. Beilstein Test

This test is to detect presence of halogens. A bright copper wire (about 5 cm long with one end embedded in a cork, which serves as a handle) is cleaned by heating to redness in a colorless bunsen flame. The material under test is touched by the hot wire and returned to the flame. A green flame, sometimes momentary, indicates the presence of halogens. It should be noted that a compounding ingredient may contain halogen, and thus give rise to a positve result.

7. Specific gravity

Some polymers have specific gravity of less than unity and hence will float on water. These are polyethylene, polyisobutylene,

Table 4.25: *Burning characteristics of rubbers*

(i) *The elastomer burns but extinguishes itself on removal from the flame*

Sl. No.	*Rubber*	*Flame colour*	*Odour*	*Other features*
1.	Chlorosulphonated poly ethylene	Bright yellow	Acrid	Grey fumes
2.	Polychloroprene	Yellow Smoky	Acrid	Acidic fumes
3.	Fluorinated rubbers	Yellow	Slight sickly	Acidic fumes

(ii) The elastomer burns and continues to burn after removal from the flame

Sl. No.	*Rubber*	*Flame colour*	*Odour*	*Other features*
1	Acrylate rubbers	Yellow smoky	Fruity	—
2	Butyl	Yellow smoky	Light sweet	
3	EPDM	Yellow with blue base	Resembles burning candle wax	—
4	Nitrile	Yellow smoky	sickly sweet	—
5	Polybutadiene or (BR)	Yellow with blue base smoky	sickly sweet	—
6	NR	Yellow, smoky	Pungent	—
7	PUR	Yellow with blue base	Acrid	—
8	SBR	Yellow, very smoky	Styrene	—
9	Polysulfide	Blue	Sulphurous	Acidic fumes
10	Silicone	Bright yellow white	None	White fumes and white residue
11	Vinylpyridine	Yellow, very smoky	Styrene	—

polypropylene (PP), styrene-butadiene rubber, butyl, ethylene-propylene, natural rubber, poly butadiene, polyisoprene, nitrile (some types), silicone and vinyl pyridine rubbers. When carrying out this test it is essential to ensure that the sample is properly wetted; it should be pushed below the surface of water in a test tube with a glass rod and then released. It is important to note that if compounding ingredients are present, the

Table 4.26: *Some typical properties of common elastomers*

Rubber	Mol. Wt. of repeating unit	ρ_c (g/cc)	ρ_a (g/cc)	ρ_c/ρ_a	Tg (K)	Tm (K)	Refractive Index (η)	Solubility parameter (δ) $cal^{1/2}/cm^{3/2}$	Solvents	Degradation Temp.* (T_d) (°C)
Cis-1,4- poly butadiene	54.1	—	0.92	—	171	285	1.516	8.1-8.6	Dipentene, benzene, xylene, decane	510
Trans-1,4- poly butadiene	54.1	—	—	—	255-263	415	—	—	Toluene, Chlorinated hydrocarbons, THF, aliphatic hydrocarbons	—
Poly sulphide rubber	156	—	1.34	—	198	—	—	9.0-9.4	Benzene	—
Poly-1-butene Rubber	56.1	0.95	0.86	1.104	228/249	379/ 415	—	7.8-8.1	Aromatic hydrocarbons, chlorinated aliphatics	429

(contd.)

Rubber Wt. of	*Mol. (g/cc) repeating unit*	ρ_c *(g/cc)*	ρ_a	ρ_c/ρ_a	*Tg (K)*	*Tm (K)*	*Refractive Index* (η)	*Solubility parameter* (δ) *cal*$^{1/2}$/*cm*$^{3/2}$	*Solvents*	*Degradation Temp.** (T_d) *(°C)*
Polyisobutylene	56.1	0.94	0.84	1.12	198	275/ 317	1.508	7.9	Aliphatic, cycloaliphatic and aromatic hydrocarbons, dioxane, THF, chlorinated hydrocarbons	
Cis-1,4 Polyisoprene (Isoprene)	68.1	1.00	0.91	1.10	200	298	—	—	Cyclohexane, benzene, toluene, xylene, Chlorinated hydrocarbons	412
Trans -1,4 (NR)	68.1	1.05	0.90	1.16	213	338	1.520	7.8-10.0	Cyclohexane	—
Polyurethane (PU)— uncrosslinked	—	—	1.23	—	222	—	—	—	Ethyl acetate, DMF, formic acid, m-cresol	—
Styrene-butadiene rubber (SBR)	158	—	0.94	—	223	—	1.52-1.55	8.4	Ethyl isobutyrate, cyclohexane, toluene	510

EPDM (Ethylene, Propylene and Butadiene)	124	—	0.85	—	235	—	—	—	Chloroform	524
Chloroprene (CR) (Neoprene)	88.5	1.36	1.24	1.09	230	383	1.558	8.2-9.25	Benzene	347
Acrylate rubber	—	—	1.17	—	238	—	—	8.4-8.8	n-butanol, chloroform, toluene, ethyl acetate	457
Nitrile rubber (NBR) (Acrylonitrile and butadiene)	107	—	0.96	—	273	—	—	—	—	518
PDMS	74.1	1.07	0.98	1.09	150	234-244	—	7.3	Chloroform, heptane, benzene, diethyl ether	706

*Temperature for maximum degradation which is obtained from TGA thermogram at a heating rate of 30°C/mm

Non-Solvents for some of the common rubbers are given in Table 4.27.

Table 4.27. *Non-solvents for some common rubbers*

Rubbers	*Non solvents*
Polybutadiene	Alcohol, lower ketones and esters, nitromethane, propionitrile, water
Polyisoprene/ natural rubber	—do—
Poly isobutylene	lower ketones, and esters, lower alcohols, nitromethane, propionitrile
Poly butylacrylate	Aliphatic hydrocarbons, diethyl ether, methanol, ethanol, ethyl acetate

specific gravity of the sample will not be the same as that of the polymer. For this reason and also because some of the polymers are available with a range of specific gravities, the determination of the actual specific gravity is not advocated here as a general identification test.

8. Odour

Some materials have pronounced odour, e.g. polysulphides and natural rubber smoked sheet, which should be compared with the odour of the known samples.

II. SOLUBILITY TEST

Solubility tests one of the simple physical tests to identify the types of rubbers. However, in many cases solubility varies considerably for different samples of the same resin and it is difficult to interpret the results. The solubility test should be carried out in a test tube. Take about 0.5 g of rubber sample and add about 5 ml of solvent, mix and occasionally shake the contents of the test tube. Observe for few minutes. Swelling or dissolution or sticky nature of polymer indicates that the polymer or the rubber is soluble in that solvent. Then refer Table 4.26 for more details.

III. ELEMENTAL ANALYSIS

The results of tests for the elements nitrogen, sulphur and halogens serve to indicate the possible nature of the unknown material. If the material is a compounded rubber (unvulcanized or vulcanized) then it is worth while extracting it by boiling a sample (about 2g) comminuted as finely as possible with scissors, with two portions (50 ml each) of methyl ethyl ketone for 15-20 min each. This may be carried out in a beaker, covered with a watch-glass, on a steam bath. The extracted sample is dried with a filter paper or, if sticky, in an oven at 100°C. This procedure, which is unsuitable for quantitative work, removes organic compounding ingredients (and their residues) which may contain nitrogen, sulphur (if the sample is vulcanized the extracted material will still contain small amounts of sulphur) and halogens.

The Lassaigne procedure of sodium fusion forms the basis of the examination for elements. A small piece of clean sodium (about 0.06 g) is placed in an ignition tube and a small quantity (about 0.03 g) of the substance to be tested is added. The tube is gently heated until reaction subsides and then strongly until the glass is red-hot. This operation can be dangerous and the mouth of the boiling tube should be covered with little Na_2Co_3. The tube is then plunged into distilled water (10 ml) contained in a pestle and mortar. The resulting mixture is heated to boiling and filtered from carbon and fragments of glass. The filtrate is divided into four equal portions which are used for the detection of elements.

Test procedures to characterize nitrogen, halogens and sulphur from sodium extraction solution can be used from part A (plastics) of chapter 4 (elemental analysis section).

Classification of rubbers on the basis of heteroatoms are given below:

Table 4.28. *Classification of Rubbers by their Heteroatoms*

Heteroatoms			
Nitrogen	*Halogens*	*Sulphur*	*Not containing N, S and halogens*
Polyurethane rubbers, Vinyl pyridine rubber, Nitrile rubber	Polychloroprene, Chlorinated butyl rubber, Hypalon, Bromobutyl rubber, Fluoro rubbers	Hypalon, Polysulphide rubber	NR, SBR, EPDM, Polybutadiene, Acrylate rubber, Silicone rubber, Butyl rubber, Ethylene-propylene rubber, Polyisoprene

IV. TRANSITION TEMPERATURES

Glass transition temperature (T_g) is a decisive factor, which can distinguish between different rubbers. Information about transition temperatures like T_g and T_m of rubbers is given in Table 4.26.

FINAL IDENTIFICATION BY CHEMICAL TESTS

Group 1: Rubbers Containing Nitrogen

1. Test for Polyurethane (PU) Rubbers

1 ml of concentrated sulphuric acid is carefully added to 4 ml of water and 0.05 g of rubber is put into the mixture. After

boiling for few minutes, polyurethane rubber disintegrates. Vinyl pyridine and nitrile rubbers are unaffected.

2. Test for Vinylpyridine Rubber

About 0.1 g sample is fused with 0.2 g potassium hydroxide in a heat resistant tube. The mixture is cooled and then shaken with 2 ml of water. The aqueous layer is poured off and just acidified with 2 ml of 5N hydrochloric acid and 5 drops of Mayer's reagent (3 g mercuric chloride and 10 g potassium iodide in 200 ml water) is added. A yellow precipitate, which disappears on warming, shows the presence of vinylpyridine rubber.

3. Test for Nitrile Rubber (NBR)

Negative results from tests 1 and 2 indicates the presence of nitrile rubber.

Group 2: Rubbers Containing Chlorine

4. Test for Polychloroprene Rubber

0.05 g of sample is boiled with 2 ml of concentrated nitric acid. Disintegration of rubber and evolution of brown fumes indicates the presence of polychloroprene. Chlorinated butyl and ethylene rubbers are not affected.

5. Test for Chlorinated Butyl Rubber

A filter paper is immersed in freshly prepared mercuric sulphate solution (mercuric oxide in 20 ml of boiling 5 N sulfuric acid and cooled before use). This paper is inserted in the mouth of an ignition tube containing 0.2g of polymer. The tube is gently heated so that the polymer just decomposes. Bright white coloration indicates the presence of butyl rubber or IIR. Presence of chlorine confirms the presence of chlorinated butyl rubber.

6. Test for Chloro Sulphonated Polyethylene (Hypalon)

Negative results from tests 4 and 5, indicates the presence of chloro sulphonated polyethylene.

7. Rubber hydrochlorides

0.1 g of the polymer is shaken with 5 ml of tetrachloroethylene. If the rubber sample is soluble, it indicates the presence of rubber hydrochlorides.

Group 3: Rubbers Containing Bromine

8. Test for Bromo Butyl Rubber

A filter paper is immersed in freshly prepared mercuric sulfate solution (mercuric oxide in 20 ml of boiling 5N sulphuric acid and cooled before use). This paper is inserted in the mouth of an ignition tube containing 0.2 g of polymer. The tube is gently heated so that the polymer just decomposes. Bright yellow coloration indicates the presence of butyl rubber or IIR. Presence of bromine in elemental analysis and positive result in this test confirms the presence of bromobutyl rubber.

Group 4: Rubbers Containing Fluorine

9. Test for Poly Fluoro Rubbers

The common fluorinated rubbers, viz., chloro trifluoro ethylene, vinylidene fluoride and perfluoro propylene-vinylidene fluoride may be distinguished by means of elemental analysis, since the former contains chlorine and latter does not.

Group 5: Rubbers Containing Sulphur

10. Test for Polysulphide Rubber

0.05 g of sample is dropped into 2 ml cold concentrated nitric acid. A vigorous reaction with evolution of brown fumes indicates the presence of polysulphide.

Group 6: Rubbers Not Containing Nitrogen, Halogens or Sulphur

About 0.05 g of sample is boiled with 2 ml concentrated nitric acid and allowed for few minutes. It is noted whether or not disintegration occurs. If disintegration occurs then the procedure outlined in section A should be followed. If disintegration doesn't occur, procedure outlined in section B should be followed.

Section A: Disintegration in Nitric Acid

11. Test for Polyisoprene (NR)

(i) 0.1 g of sample is heated in a water bath with 5 ml chromic acid solution (prepared by dissolving 2 g of chromic trioxide in 5 ml water and added 1.5ml of concentrated sulphuric acid).

If NR or polyisoprene is present, acetic acid odour is appeared after few minutes.

(ii) About 0.05 g of extracted rubber sample is heated in water bath with 5ml of bromine in carbon tetrachloride (10% VN) until no bromine remains. A solution of 5 ml of phenol in carbon tetrachloride (10% VN) is added and heating is continued until only a small amount of about 0.5ml of liquid remains. In the presence of polyisoprene, the solution becomes violet. This along with the first test confirms the presence of polyisoprene. This test alone is not sufficient because even IIR gives a positive result to this test.

12. Test for SBR

Conduct styrene test as given under plastic section (test for PS). If the result is positive then the given rubber is SBR.

13. Test for Polybutadiene (BR)

Negative results from tests 7 and 11 indicate the presence of BR.

Section B: Non-disintegration in Nitric Acid

14. Test for Acrylate Rubbers

About 0.1g of sample is boiled with 5 ml of acetic acid dissolution or swelling after few minutes indicates the presence of acrylate rubber.

15. Test for Butyl Rubber (IIR)

Conduct tests 8 and 12, if they are positive, then the given rubber is butyl.

16. Test for Silicone Rubber

(i) About 0.2 g of rubber is ignited on a spatula and the flame is allowed to pass on to the center of a small watch glass. In the presence of silicone, the glass becomes coated with a white deposit of silica. The glass is cooled and the deposit is treated with little hydrofluoric acid. Silica dissolves and evaporation on a steam bath leaves no residue.

(ii) 0.5 g of the sammple, 1g of dry sodium carbonate and 0.1g of sodium peroxide are heated in small platinum or nickel crucible.

(a) After cooling, the melt is dissolved in dilute 1:1 hydrochloric acid, the resulting solution is transferred to a porcelain dish and evaporated to dryness. The solid residue is treated with few ml of dilute hydrochloric acid, washed once or twice with water, moistened on the filter paper with 0.1% methylene blue water. The residue on the filter paper turns intense blue due to absorption of the dye by silica, hence confirming the presence of silicones.

(b) After cooling dissolve the rubber in a few drops of water, bring it to boil quickly, neutralize and slightly acidify with dilute nitric acid. To this one drop of molybdate solution (dissolve 30 g ammonia molybdate in 60 ml of hot water, cool and make up to 100 ml by adding water. Then add a solution of 10 g ammonium sulphate in 100 ml of 55% nitric acid) is added and allowed to stand for 24 hours. The supernatant liquid is removed by suction decantation, well sealed and kept in dark. Then it is heated to boiling, cooled and one drop of benzidine solution (50 mg benzidine dissolved in 10 ml of 50% acetic acid and 90 ml water) and one drop of saturated aqueous sodium acetate is added to it. A blue colour indicates the presence of silicones.

17. Test for EPDM/EPR

Negative results from tests 14, 15 and 16 indicates the presence of ethylene propylene diene rubber (EPDM/EPR).

18. General Identification Reaction

Burchfield color reaction is useful to differentiate between different types of rubbers (see Table 4.29). Heat 0.5 g of the sample in a test tube. Pass the pyrolysis vapours into 1.5 ml of reagent (reagent-dissolve 1 g p-dimethyl amino benzaldehyde and 0.01 g hydroquinone by heating them gently in 100 ml methanol. Then react the solution with 5 ml conc HCl and 10 ml of ethylene glycol. This reagent can be kept for several months in a brown bottle). Observe the colour, dilute the solution with 5 ml of methanol and boil for 3 min.

SPOT TEST

Introduction: The identification of rubber has been achieved in the past through the use of colour reactions as well as with the help of various analytical techniques. The color reaction

Table 4.29: *Burchfield colour reaction to distinguish between elastomers*

Elastomer	*On contact of the pyrolysis vapour with reagent*	*After addition of methanol and boiling for 3 min*
None (Blank tests)	Yellowish	Yellowish
Natural rubber (polyisoprene)	Yellow-brown	Green-violet-blue
Poly butadiene	Light-green	Blue-green
Butyl rubber	Yellow	Yellow-brown to weakly violet-blue
Styrene-butadiene copolymers	Yellow-green	Green
Butadiene-acrylonitrile copolymers	Orange red	Red-red brown
Polychloroprene	Yellow-green	Yellowish - green
Silicone rubber	Yellow	Yellow
Polyurethane elastomers	Yellow	Yellow

has been used by several workers for differentiating between elastomers, owing to the almost unlimited range fo this reaction with consequent color formation.

Spot testing, on filter paper, has been evolved as a micro test for the identification of the more common rubbers used in the rubber industry. This section deals with the identification of rubber by two spot test methods, namely, (A) using filter paper strips and (B) Test in solutions.

A. Using Filter Paper Strips

(a) CR and NBR spot test

(i) **Preparation of test paper:** About 2 g of copper acetate and 0.25 g of metanil yellow is dissolved in 500 ml of methanol and a filter paper is impregnated in it.

(ii) **Wetting solution**: About 25 g of benzidine hydrochloride is dissolved in a mixture of 500 ml of water and 500 ml methanol, 10 ml of an aqueous solution (1 %) of hydroquinone is added and kept in a brown bottle.

(b) Butyl Rubber (IIR) spot test

(i) **Preparation of test paper:** Blank filter paper strips are used.

(ii) **Wetting solution:** 5 g of mercuric oxide is added to a mixture of 15 ml H_2SO_4 and 80 ml water and boiled. The heating is continued until the oxide is dissolved. The solution is then cooled and diluted with 100 ml of diethyl ether.

(c) SBR and IR spot test

(i) **Preparation of test paper:** Filter paper is impregnated in a solution of 3 g of p-dimethylamino benzaldehyde and 0.05 g of hydroquinone in 100 ml of diethyl ether.

(ii) **Wetting solution:** About 30 g of trichloro acetic acid is dissolved in isopropanol and further diluted with 100 ml of isopropanol.

Procedure for the Spot Test

Add a small rubber (approximately 100 mg) to a pyrolysis tube, gripping the upper end of the tube with a clip of tongs At the open end of the tube place a strip of wetted filter paper with wetting solution. Heat the test tube in the flame of a Bunsen burner that has been reduced to minimum heat, taking care to point the open end of the tube away from the face. Depending on the reaction of the escaping vapours (decomposition product) of rubber with strip of wetted filter paper it is possible to distinguish between different rubbers. Details of the spot test result is given in Table 4.30.

Table 4.30: *Observation and Inference of the spot test*

Rubber	*CR-NBR test*	*IIR test*	*SBR-NR-IR test*
CR	Red	—	Green
NBR	Green	Pale brown	Yellow green
SBR	—	Brown	Blue green
NR or IR	—	Brown	Blue
IIR	—	Yellow	Pale lavender

B. Test In Solutions

(a) Spot test solution I: 1 g of p-dimethylamino benzaldehyde and 0.01 g of hydroquinone are added to 100 ml of methanol and made to dissolve in 5 ml of conc. HCl + 10 ml of ethylene glycol and specific gravity is adjusted to 0.851 by adding ethylene glycol.

(b) Spot test solution II: 2 g of sodium citrate, 10.2 g of citric acid and 0.03 g of metanil yellow are dissolved in 500 ml of water.

Procedure: 0.5 g of sample is gently heated and the vapors are passed through solutions I and II. The colour changes after the chemical reactions of pyrolysed products of rubbers with reagent solutions I and II is given in Table 4.31.

Table 4.31: *Observations and Inference of the Spot Test solutions I and II*

Rubber	*Solution I*		*Solution II*
	Initial	*Final*	
CR	Yellow	Pale yellow	Red
NBR	Orange-red	Red	Green
SBR	Yellow-green	Green	Green
NR/IR	Brown	Violet-blue	Green
IIR	Yellow	Pale blue-green	Green

MISCELLANEOUS PROPERTIES OF RUBBERS

Evaluation of Specific Gravity and Specific Volume of Compounded Rubber

The specific gravity is the ratio between the weight of a unit volume of the substance and to the weight of the same volume of water at a given temperature, usually at $4^{o}C$. The specific volume (which is reciprocal of density) represents the volume in CC/g. Below given steps are followed to evaluate the specific gravity of a compounded rubber.

(1) Express the weight of each ingredient in grams.

(2) Divide the weight of each ingredient by its specific gravity and add the results together to give total volume. Divide the sum of total weight of the compound by the total volume to get the specific gravity of the compounded rubber (see Table 4.32).

Table 4.32: *Illustration to calculate specific gravity of compounded and unvulcanized rubber*

Ingredient	*Weight (g)*	*Sp. Gr*	*Volume = Weight/Sp. Gr. (CC)*
BMR grade rubber	100	0.92	108.90
Zinc oxide	5	5.55	0.90
Stearic acid	3	0.85	3.53
Sulphur	3	2.10	1.43
Accelerator	0.85	1.45	0.59
Total	**111.85**		**115.35**

$$\text{Specific gravity} = \frac{111.85}{115.35} = 0.97$$

The method of calculation shown above gives the density of the uncured stock. The density of the cured stock is somewhat higher due to decrease in free volume and void content, that accompanies the combination of rubber with sulphur. For small sulphur ratios, this volume decrease can be taken into account by using the fraction value of 6 for the density of sulphur. Table 4.33 gives the specific gravity of some important rubber compounding ingredients.

Table 4.33: *Specific gravity of rubber compounding ingredients*

Ingredient	*Sp. gr.*	*Ingredient*	*Sp. Gr.*
Aluminum flake	2.50	Lime	2.19
Aluminum silicate	2.10	Litharge	9.30
Aluminum carbonate	1.59	Lithopone	4.30
Antimony trioxide	5.40	Magnesia	3.60
Asbestos	2.70	Mica	2.80
Barytes	4.50	Mineral oil	0.91
Bentonite	2.50	MRX	1.01
Bitumen	1.04	Oleic acid	0.91
Blanc fixe	4.30	Palm oil	0.75-0.85
Calcium silicate	2.10	Paraffin oil	0.80
Carbon black	1.80	Paraffin wax	0.90
Camauba wax	0.99	Petroleum jelly	0.90
Casein	1.26	Pine tar	1.08
Ceresine wax	0.92	Rosin	1.07

(contd.)

Ingredient	*Sp. gr.*	*Ingredient*	*Sp. Gr.*
China clay	2.50	Rosin oil	0.98
Cork	0.24	Shellac	1.15
Cotton fibre	1.05	Silica	1.95
Cotton seed oil	0.93	Stearic acid	0.85
French chalk	2.80	Sulphur	2.05
Glue	1.27	Talc	2.68
Graphite	2.40	Titanium-dioxide	3.90
Ground glass	2.50	Ultramarine	2.50
Halloysitic clay	2.35	Vermilion	8.20
Indene-cumarone	1.09	Whiting powder	2.65
Kerosene	0.75-0.85	Zinc carbonate	3.85
Kieselguhr	220	Zinc oxide	5.55
Lanolin	1.08	Zinc stearate	1.10
Lead oxide	9.40		

Table 4.34: *Properties and applications of some important commercial rubbers*

Rubber	*Vulcanizating agent*	*Properties (stretching crystallization)*	*Gum strength*	*Applications*
NR	Sulphur	Good	Good	Tyres, balloons
BR	Sulphur	Media	Poor to fair	Tyres, play balls
SBR	Sulphur	Poor	Poor	Tyres, carpet under lay, shoe-soles
CR	MgO, ZnO	Good	Good	Joint seals, conveyor belts, bridge bearings, electric connections, tarpaulins
IIR	Sulphur	Good	Good	Inner tubes, bags, bladders, water proof membranes
NBR	Sulphur	Poor	Poor	Gaskets, petrol hoses
EPDM	Sulphur	Poor	Poor	Inner tubes, bags, bladders, water proof sheets, cable insulation.

(*contd.*)

Rubber	*Vulcanizating agent*	*Properties (stretching crystalli-zation)*	*Gum strength*	*Applications*
PUR	Di-isocyanates	Medium	Good	Shoe-soles, seals, car bumpers, tubing
Sulphide	ZnO	Fair	Poor	Sealant, printer roller
Acrylic	Peroxides	Fair	Good	Ignition wire sheathing hoses, bellows.
Silicone	Peroxides	Poor	Poor	Shaft sealing rings, spark plug caps, radiator hoses, electrical insulation seals

Table 4.35: *Trade names and manufacturers of some of the commonly used elastomers*

Elastomers	*Trade name*	*Manufacturer*
Polyisoprene	Natsyn	Goodyear
	Afprene	Carbochem
	Europrene SOL	Enichem
Butadiene Rubber (BR)	Budene	Goodyear
	Ameripol CB	B F Goodrich
	Afdene BR	Carbochem
	Duragene	General Tire
Styrene–Butadiene Rubber (SBR)	Ameripol	B F Goodrich
	Afsol	Carbochem
	Kratex	Chemopetrol
Butadiene–Acrylonitrile Rubber (NBR)	Chemigum	Goodyear
	Hycar	BF Goodrich
	Europrene N	Enichem
	Buna N	VEB Buna
Nitrile Rubber–PVC Blends (NBR/PVC)	Hycar	BF Goodrich
	Europrene S Ozo	Enichem
Chloroprene Rubber (CR)	Neoprene	Du Pont
	Baypren	Bayer
	Skyprene	Toya Soda

(*contd.*)

Elastomers	*Trade name*	*Manufacturer*
Isobutylene–Isoprene Rubber (IIR)	Esso Butyl Polysar Butyl	Exxon Polysar
Ethylene Propylene Rubber (EPM, EPDM)	Vistalon Epcar Nordel	Exxon BF Goodrich Du Pont
Ethylene Vinylacetate Rubber	Escorene Ultra Ultrathene	Exxon USI Chemicals
Chlorosulphonated Polyethylene (CSM)	CS PE Hypalon	China Du Pont
Acrylic Rubber (ACM)	Acralen Europrene AR HyTemp	Bayer (Latex) Enichem BF Goodrich
Epichlorohydrin Rubber (CO, ECO, ETTER)	Hydrin Epichloromer	BF Goodrich Osaka Soda
Fluorinated Rubber (TFE/PP resp.)	Aflas, Kalrez	Asahi Glass, 3M Du Pont
Polysiloxane Rubber	Silastic Tufsel	Dow Corning General Electric
Fluorosilicone Rubber	FSE VP-R	General Electric Wacker
Polyurethane Rubber (AU, EU)	Adiprene Urepan Vibrathane	Du Pont Bayer Uniroyal
Poly sulfide Rubber (TM)	Thiokol Thiokol LP	Thiokol
Ethylene Acrylate Rubber (EAM)	Vamac	Du Pont
Chlorinated Polyethylene (CM)	Bayer CM Dow CPE	Bayer Dow Corning

Table 4.36: *Typical properties of commercial elastomers*

Commercial Elastomers	*Tensile Strength (psi)*	*Elongation (%)*	*Modulus (psi, 300 -400% elongation)*	*Abrasion resistance*	*Lower use temperature (°C)*	*Ozone resistance*	*Oil resistance*	*Water swelling resistance*
Natural rubber	4000	700	2500	Fair	–50	Poor	Poor	Excellent
SBR	3500	700	2500	Good	–50	Fair	Poor	Good
Butyl rubber	3000	700	1000	Good	–50	Excellent	Poor	Excellent
EPDM	3000	300	–	–	–40	Excellent	Poor	Excellent
Fluorinated rubbers	2400	400	250	–	–40	Excellent	Excellent	Excellent
Neoprene rubber	4000	700	1000	Good	–50	Good	Fair	Good
Nitrile rubber (NBR)	4000	600	1500	–	–50	Fair	Excellent	Excellent
Polybutadiene (*cis*-1,4)	3000	700	–	–	–60	Poor	Poor	Excellent
Polyisoprene (*cis*-1,4	4000	750	2500	Fair	–50	Poor	Poor	Excellent
Polysulphide	1200	400	1400	Poor	–50	Good	Excellent	Excellent
Silicone rubber	1500	800	–	Poor	–80	Excellent	Fair	Excellent
Urethane	8000	700	1200	Excellent	–50	Excellent	Excellent	Fair
Acrylate	2200	400	–	Good	–20	Excellent	Excellent	Poor
Chlorosulfonated Polyethylene	2800	500	–	–	–50	Excellent	Good	Excellent
Epichlorohydrin	2500	400	–	–	–45	Good	Good	Good

Table 4.37: *Standard tests, methods and units*

Sl. No.	*Property*	*Applicable Standards*	*Unit of Expression*
1.	Density	ASTM D 792, ISO 1183, ISO 1675 JIS K 7112 BS 2782 P5 M 509 A DIN 53479	g/cc
2.	Density By Gradient Column Technique	ASTM D 1505 ISO 1183 JIS K 7112 BS 2782 P5 M 509 B DIN 53479	g/cc
3.	Apparent Density For Powder Material	ASTM D 1895 ISO 60 BS 2782 P5 M 501 A DIN 53468	g/cc
4.	Apparent Density For Rigid Foam	ASTM D 1622 ISO 845 JIS 7222 DIN 53420	kg/cc
5.	Water Absorption	ASTM D 570 ISO 62 JIS K 7209 BS 2782 P5 M 502 F DIN 53495	% mg
6.	Sieve Analysis (Particle Size)	ASTM D 1921 ISO 4576 JIS K 7219 DIN 53477	mg %
7.	Chemical Resistance	ASTM D 543 ISO 175 JIS K 7114 BS 2782 P5 M 505 A	%
8.	Gas Permeability	ASTM D 1434 ISO 2556 BS 2782 P5 M 514 A DIN 53380	cc/m^2 24 hr atm, ml/m^2 24 hr atms
9.	Melting Point	ASTM D 2117 ISO 3146 JIS K 6810	°C, °F

(*contd.*)

Sl. No.	*Property*	*Applicable Standards*	*Unit of Expression*
		BS 2782 P1 M 103 and 103D DIN 53181	
10.	Dilute Solution Viscosity	ASTM D 2857 ISO 1628/1 BS 2782 P4 M 404 A	dl/g, Stroke, centistroke.
11.	Brookfield Viscosity	ASTM D 1824 ISO 2555	cPa
12.	Viscosity By Rotational Viscometer	ASTM D 3219	Pascals
13.	Carbon Black Content	ASTM D 1603 BS 2782 P4 405 A	%
14.	Elevated Aging Using Oven	ASTM D 1870 ISO 1637 JIS K 7212	%
15.	Water Vapour Transmission Rate (WVTR)	ASTM E 96 ISO 1663 JIS Z 0208 BS 2782 P5 M 513A DIN 53122 P1	ml/100 in^2
16.	Acetone Extraction	ISO 59 BS 2782 P4 M 401 A	%
17.	Tensile Strength	ASTM D 638 ISO 527 JIS K 7113	kg/cm^2, N/mm^2, MPa, PSI
18.	Flexural Strength	ASTM D790 ISO 178 JIS K 7203 BS 2782 P3 M 304 DIN 53452	kg/cm^2, N/mm^2, MPa, PSI
19.	Compressive Strength	ASTM D 695 ISO 604 JIS K 7208 BS 2782 P3 M 303 A DIN 53454	kg/cm^2, N/mm^2, MPa, PSI
20.	Shear Strength	ASTM D 732 JIS K 7214 BS 2782 P3 M 305	kg/cm^2, N/mm^2, MPa, PSI

(*contd.*)

Sl. No.	*Property*	*Applicable Standards*	*Unit of expression*
21.	Izod Impact Strength	ASTM D 256 A ISO 180 JIS K 7110 BS 2782 P3 M 306A	kg cm/cm, J/m, ft. lb/in
22.	Co-efficient of Friction	ASTM D 1894 ISO 8295 BS 2782 P3 M 311 A DIN 53375	Number
23.	Thermal Conductivity	ASTM D 1894 ISO 2581 JIS K 9511	K. cal/h.m°C, Btu. In/ h.ft^2. °F, W/mK
24.	Linear Thermal Expansion	ASTM D 696	mm/mm/°C, 10^{-5} K^{-1}
25.	Heat Deflection Temperature	ASTM D 648 ISO 75 BS 2782 P1M 102 G and H	°C, °F
26.	Vicat Softening Point	ASTM D 1525	°C, °F
27.	Cup Flow Test	ASTM D 731 BS 2782 P1 M 105 B DIN 53465	Sec
28.	Melt Flow Index	ASTM D 1238 ISO 1133 BS 2782 P1 M 105 C	g/10 min
29.	Dielectric Strength	ASTM D 149 ISO 1325 DIN 53483	KV/mm
30.	Dielectric Constant	ASTM D 150	Number at 60 Hz, 50 Hz
31.	Volume and Surface Resistivity	ASTM D 257 ISO 3915	Ohm-cm, Ohm
32.	Refractive Index	ASTM D 542 ISO 486 DIN 53491	Number
33.	Haze and Luminous Transmittance	ASTM D 1003 BS 2782	%
34.	Color Index	ASTM D 1925	Number
35.	Clarity	ASTM D 1746	%

EXERCISE

1. Distinguish between
   (i) PET & PBT (ii) LDPE & HDPE (iii) Nylon 6 & Nylon 66 (iv) PS, SAN & ABS
2. Which of the following polymers are natural and which are synthetic? Cellulose, polystyrene, chitosan, polysaccharides, polyvinyl alcohol, natural rubber, silk, pullulan, lignin, nylon, protein, epoxies, phenol-formaldehyde.
3. Identify which of the following polymers are condensation polymers and which are addition polymers: PET, Nylon, PMMA, PC, PAN, PEEK, LLDPE, PS, PVAc, UF, PU, PP, NBR, BR.
4. Which are the polymers burns with sooty flame? Why?
5. Why nitrogen and halogen containing polymers will not burn easily?
6. Which polymers yield more ash content? Why?
7. Name a few nitrogen, halogen and sulphur containing polymers
8. Give few examples for thermoplastic and thermoset.
9. Give some examples for rubbers and fibers.
10. Name few engineering plastics.
11. In each set which polymer has higher melting point and why?
    (i) PET & PBT (ii) LDPE & HDPE (iii) Nylon 6 & Nylon 66
12. Write the structural formula for the following polymers: (i) PS (ii) PMMA (iii) PC (iv) PET and (v) PAN
13. How do you check the acidic or basic nature of the polymer solution.
14. Explain Beilstein test.
15. Name the solvents for the following polymers (eg.- PS, PMMA, NR, PVC, Nylon 6, PAN, etc)
16. Why do you get different coloured precipitation while detecting halogens?
17. Define the term softening point. Give its importance.
18. Explain the terms T_g and T_m with examples.
19. Define the term MFI. What is its significance?
20. Name the factors, which affects the solubility of the polymer.
21. What is the significance of burning tests in the identification of polymers?
22. Is it possible to differentiate the polymers on the basis of density? Explain.

23. Mention the factor affects on the density of the polymers.
24. Define the term tacticity. Write down the different structure of PP on the basis of tacticity.
25. How do you differentiate isotactic PP from atactic PP?
26. How do you identify the different stereo isomer polymers by simple physical methods?
27. Which polymer has the lowest density?
28. Why hetero polymers (condensation polymer products) possess high density and high melting point?
29. Name some transparent polymers. Why they are transparent?
30. What do you understand by the terms raw rubber, creep rubber, vulcanized rubber, accelerator and anti-oxidant.
31. Write the structural formulae of the following synthetic rubbers, SBR, neoprene, polyisoprene, NBR, chloroprene rubber, EPDM, EPR, IIR.

References

Allen, Sir Geoffrey and Bevington, J.C, eds., Polymer Characterization', Vol. I of *Comprehensive Polymer Science*, Pergamon Press, England, 1988.

Ash, M.B. and Ash, I.A. eds., *Handbook of Plastic Compounds, Elastomers and Resin*, VCH Pub. Inc., New York, 1992.

Bandrup J. and Immergu, E.H. eds., *Polymer Handbook*, 3rd edn., Wiley-Interscience, New York, 1989.

Bark, L.S. and Allen, N. S. eds., *Analysis of Polymer Systems*, Applied Science Publishers Ltd., London, 1982.

Bicerano, J., *Prediction of Polymer Properties*, Plastics Engineering Series, 27, Marcel Dekker, Inc., New York, 1993.

Braun, D.: *Simple Methods for Identification of Plastics*, 2nd edn., Hanser, Munich, 1986.

Brydson, J.A., *Plastic Materials*, 6th edn., Butterworths Heinemann Ltd., Oxford), 1995.

Brydson, J.A., *Handbook for Plastics Processors*, Heinemann Newner, Oxford, 1990.

Cassidy, P.E., '*Thermally Stable Polymers, Synthesis and Properties*, Marcel Dekker, Inc., New York, 1980.

Chemical Resistance Data Sheets, Vol. I and II, RAPRA Pub., 1993.

Cheremisinoff, N.P. edn., *Handbook of Engineering Polymeric Materials*, Marcel Dekker, Inc., New York, 1997.

Cheremisinoff, N.P. and Cheremisinoff, P.N. eds., *Handbook of Advanced Materials Testing*, Marcel Dekker, Inc., New York, 1994.

Collins, E.A., Bares, J. and Billmeyer, Jr., F.W.: *Experiments in Polymer Science*, Wiley-Intersceince, New York, 1973.

Crompton, T.R.; *Chemical Analysis of Additives in Plastics*, 2nd edn., Pergamon, Oxford, New York, 1977.

F.W., Billmeyer, J.R., *Text book of Polymer Science*, (3rd edn.) Wiley Interscience Publication, New York, 2002.

Fava, R.A. edn., *Methods of Experimental Physics*, Vol. 16, Part B, *Polymers: Crystal Structure and Morphology*, Academic Press, Inc., New York, 1980.

Griffin, L.O., *Physical Constants of Linear Homopolymers*, Springer, New York, 1968.

Harper, C.A., Editor-in-chief, *Handbook of Plastics and Elasomers*, McGraw-Hill Book Co., New York, 1975.

Haslam, J., Willis, H.A. and Squirrel, D.C.M.: *Identification and Analysis of Plastics*, 2nd Ed., Butterworth, London 1972; Paperback Reprint Edition, Heyden & Son, London and Philadelphia, 1980.

Hofman, W., *Rubber Technology Handbook*, Hanswer Publishers, New York, 1989.

Hunt, B.J. and James, M.I. eds, *Polymer Characterization*, Chapman and Hall, New York, 1993.

Imanishi, Y. (ed), *New Polymers*, Huthig and Wepf, Germany, 1992.

Kämpf, G.: *Characterization of Plastics by Physical Methods-Experimental Techniques and Practical Applications*, Hanser, Munich, Vienna, 1986.

Krause, A., Lange, A. Ezrin, M., *Plastics Analysis Guide–Chemical and Instrumental Methods*, Hanser, Munich, Vienna, 1983.

McCaffery, E.L.: *Laboratory Preparation of Macromolecular Chemistry*, McGraw-Hill, New York, 1970.

Mitchell, J.Jr. (Ed.): *Applied Polymer Analysis and Characterization*, Hanser, Munich, Vienna, 1987.

Pinner, S.H.: *A Practical Course in Polymer Chemistry*, Pergamon Press Ltd., London, 1961.

Redfarn, C.A. and Bedford, J.: *Experimental Plastics*, Hiffe, London, 1960.

Saunders, K.J.: *The Identification of Plastics and Rubbers*, Chapman and Hall Ltd., London, 1966.

Schröder, E.: *Guide to Polymer Characterization*, Hanser, Munich, Vienna, 1988.

Sorensen, W. and Campbell, T.W.: *Preparative Methods of Polymer Chemistry*, Wiley-Interscience, New York, 2nd Ed., 1968.

Van Krevelen, D.W., *Properties of Polymers*, Elsevier Science Publishers, Amsterdam, 1990.

Van Krevelen, D.W. and Hoftyzer, P.J.: *Properties of Polymers-Correlations with Chemical Structures*, Elsevier Publishing Co., Amsterdam, 1972.

Wake, W.C.: *Analysis of Rubbers and Rubber-like Polymers*, 2nd Ed., Maclaren, London, 1969.

Appendices

Appendix I Abbreviations for Polymers — Plastics and Rubbers

Appendix II Glossary of Terms

Appendix III Conversion Tables

APPENDIX I

Abbreviations for Polymers for Plastics and Rubbers

Plastics

ABS	Acrylonitrile-butadiene-styrene polymer
ACS	Acrylonitrile chlorinated polyethylene and styrene
AES	Acrylonitrile-ethylene-propylene-styrene polymer
AMC	Alkyd moulding powder
AMMA	Acrylonitrile-methyl methacrylate copolymer
APP	Atactic polypropylene
ASA	Acrylonitrile-styrene-acrylate terpolymer
BOPP	Biaxially oriented polypropylene
BS	Butadiene-styrene copolymer
CA	Cellulose acetate
CAB	Cellulose acetate-butyrate
CAP	Cellulose acetate-propionate
CF	Cresol-formaldehyde resin
CMC	Carboxymethyl cellulose
CN	Cellulose nitrate
CP	Cellulose propionate
CPE	Chlorinated polyethylene
CPVC	Chlorinated polyvinyl chloride
CS	Casein
CSM	Chlorosulphonated polyethylene
CTA	Cellulose triacetate
CTFE	Poly (mono chlorotrifluoroethylene)
EAA	Ethylene-acrylic acid copolymer
EC	Ethyl cellulose
ECB	Ethylene copolymer blends with bitumen

ECTFE	Ethylene-chlorotrifluoroethylene copolymer
EEA	Ethylene-ethyl acrylate copolymer
EMA	Ethylene-methacrylic acid copolymer or Ethylene-maleic anhydride copolymer
EP	Epoxy resin
EPE	Epoxy resin ester
EPS	Expanded Polystyrene
EVA	Ethylene-vinyl acetate copolymer
EVE	Ethylene-vinyl ether copolymer
FRP	Fibre reinforced plastic
GPPS	General purpose polystyrene
HDPE	High density polyethylene
HEC	Hydroxyethyl cellulose
HIPS	High impact polystyrene
HMWHDPE	High molecular weight high density polyethylene
IPN	Interpenetrating polymer network
LCP	Liquid crystalline polymer
LDPE	Low density polyethylene
LLDPE	Linear low density polyethylene
MABS	Methyl methacrylate-acrylonitrile-butadiene-styrene
MBS	Methyl methacrylate-butadiene-styrene terpolymer
MC	Methyl cellulose
MDPE	Medium density polyethylene
MF	Melamine-formaldehyde resin
MPF	Melamine-phenol-formaldehyde resin
NC	Nitrocellulose
OPP	Oriented polypropylene
PA	Polyamide
PAA	Polyacrylic acid
PAI	Polyamide-imide
PAMS	Poly alpha-methyl styrene
PAN	Polyacrylonitrile
PBMA	Poly n-butyl methacrylate

PBT	Polybutylene terephthalate
PC	Polycarbonate
PCD	Polycarbodimide
PCTFE	Poly chloro tri fluoro ethylene
PDAP	Poly diallyl phthalate
PDMS	Poly dimethyl siloxane
PEA	Poly ethyl acrylate
PEEK	Polyether ether ketone
PEI	Poly etherimide
PEK	Polyether ketone
PEO	Poly ethylene oxide
PES	Polyether sulphone
PET	Polyethylene terephthalate
PETG	Polyethylene terephthalate (Glycol modified)
PF	Phenol formaldehyde
PI	Polyimide
PMA	Polymethyl acrylate
PMP	Polymethyl pentene
PMMA	Polymethyl methacrylate
POM	Polyoxymethylene
PP	Polypropylene
PPE	Poly phenylene ether
PPO	Poly phenylene oxide
PPS	Polyphenylene sulphide
PPT	Polypropylene terephthalate
PS	Polystyrene
PSU	Polysulphone
PTFE	Polytetrafluoro ethylene
PTMT	Polytetramethylene terephthalate
PU	Polyurethane
PVA	Poly vinyl alcohol
PVAc	Polyvinyl acetate
PVC	Polyvinyl chloride
PVDC	Polyvinylidene chloride

PVDF	Polyvinylidene fluoride
PVF	Polyvinyl fluoride
PVP	Poly (N-vinyl pyrrolidene)
RF	Resorcinol-formaldehyde resin
RPVC	Rigid polyvinyl chloride
SAN	Styrene-acrylonitrile copolymer
SHIPS	Super high impact polystyrene
SIS	Styrene-isoprene block copolymer
SMA	Styrene-maleic anhydride copolymer
SMS	Styrene α-methyl styrene copolymer
TPR	1,5-trans polypentanemer
TPU	Thermoplastic polyurethane
UF	Urea-formaldehyde
UHMWPE	Ultra high molecular weight polyethylene
USP	Unsaturated polyester resin
UPVC	Unplasticized polyvinyl chloride
VLDPE	Very low density polyethylene
VC/EC	Vinylchloride-ethylene copolymer
VC/EVAc	Vinylchloride-ethylene vinyl acetate copolymer
VC/MA	Vinylchloride-methyl acrylate copolymer
VC/MMA	Vinylchloride-methyl methacrylate copolymer
VC/OA	Vinylchloride-octyl acrylate copolymer
VC/VAC	Vinylchloride-vinyl acetate copolymer
VC/VDC	Vinylchloride-vinylidene chloride copolymer
XLPE	Cross linked polyethylene
XPS	Expandable or expanded polystyrene

FOR RUBBERS (BASED ON ISO RECOMMENDATION AND ASTM D1418)

ABR	Acrylate-butadiene rubber
ACM	Acrylate rubber
ANMA	Acrylonitrile-methyl acrylate rubber
BIIR	Bromo-isobutene-isoprene rubber (brominated butyl rubber)
BR	cis-1, 4-Butadiene rubber (cis-1, 4-polybutadiene)

BS	Butadiene-Styrene copolymer
CHC	Epichlorohydrin-ethylene oxide rubber
CHR	Epichlorohydrin rubber
CIIR	Chloro-isobutene-isoprene rubber (chlorinated butyl rubber)
CPE	Chlorinated polyethylene
CNR	Carboxynitroso rubber (tetra fluoro ethylene-trifluoro nitroso methane- unsaturated monomer terpolymer)
CO/CHR	Poly [(chloromethyl) oxirane]; Epichlorohydrin rubber
CR	Chloroprene rubber
ENR	Epoxidized natural rubber
EPDM	Ethylene-propylene non-conjugated diene terpolymer
EPT	Ethylene-propylene-diene terpolymer
EPR	Ethylene-propylene rubber
EVA	Ethylene-vinyl acetate rubber
PU	Polyether urethanes
FE	Fluorine-containing elastomer
FEP	Tetrafluoroethylene-hexafluoropropylene rubber (PEEP)
FSI	Fluorinated silicone rubber
HSBS	Hydrogenated SBS
POR	Polypropylene oxide rubbers
BR (GR-I)	Butyl rubber (former US acronym) (also IIR, PIBI)
GR-N	Nitrile rubber (former US acronym) (NBR)
SBR (GR-S)	Styrene-butadiene rubber (former US acronym) (PBS)
HNBR	Hydrogenated nitrile rubber
IIR	Isobutene-isoprene rubber; butyl rubber
NIR	Synthetic cis-1, 4-polyisoprene
NBR	Nitrile rubber (Acrylonitrile-butadiene rubber)
NCR	Acrylonitrile-chloroprene rubber
NIR	Acrylonitrile-isoprene rubber

NR	Natural rubber (cis-1, 4-polyisoprene)
OER	Oil extended rubber
OPR	Propylene oxide rubber
TPR, PEBA	Thermoplastic polyamide rubber, polyether block amide
PB	Poly (1-butene)
PIB	Polyisobutylene
PDMS	Polydimethyl siloxane
PO	Polyoctenamer
S-EPDM	Sulphonated ethylene-propylene-diene terepolymer
SCR	Styrene-chloroprene rubber
SBR	Styrene-butadiene rubber (PSB)
SBS	Styrene-butadiene-styrene triblock copolymer
SIR	Standard Indonesian rubber or styrene-isoprene rubber.
SIS	Styrene-isoprene-styrene triblock copolymer
SMR	Standard Malaysian rubber
TPE	Thermoplastic elastomer
TPO	Thermoplastic polyolefine rubber
TPU	Thermoplastic polyurethane

APPENDIX II

Glossary of Terms

Accelerator : A chemical used to speed up a curing reaction. For example cobalt naphthenate is used to acelerate the curing reaction of polyester resin. Acelearator is often referred as promoter.

Aging: The permanent change in properties of a material with time under specified conditions.

American National Standards Institute (ANSI), New York: ANSI is a non-profit organization consisting of members from commerce, industry, professionals, trade consumers, labor organizations, and government.

American Society for Testing and Materials (ASTM) : ASTM is a scientific and technical organization formed for "the development of standards on characteristics and performance of materials, products, systems and services, and the promotion of related knowledge". ASTM is the world's largest source of voluntary consensus standards.

Amorphous polymers: Polymeric materials that have no definite order or crystallinity. The polymer molecules are arranged in completely random fashion, e.g. PVC, PC, PS, NR, etc.

Anti-oxidant: Compounding material used to retard deterioration caused by oxidation

Apparent density : The weight in air of a unit volume of a material before molding.

Arc resistance: Time required for a given electrical current to render the surface of a material conductive because of carbonization by the arc flames (ASTM D 495, vol. 08.01)

Atactic polymer: Polymer in which there is a random arrangement of pendant groups on each side of the chain.

BIS : Bureau of Indian Standards

BS : British Standard

Bulk density : The weight per unit volume of a material including voids inherent in material as tested.

Bulk factor : The ratio of the volume of a given mass of molding material to its volume in the molded form.

Capacitance: That property of a system of conductors or dielectrics, which permits the storage of electricity when potential difference exists between the conductors. This value is expressed as the ratio of quantity of electricity to a potential difference. A capacitance value is always positive.

Cellular material: A generic term for materials containing many cells (either open, closed or both) dispersed throughout the mass.

Chain transfer: A reaction in which a free radical abstracts an atom or group of atoms from a solvent, initiator, monomer or polymer.

Chalking (rubbers): The formation of a powdery residue on the surface of a rubber, resulting from surface degradation.

Coagulation (rubber latex): An irreversible agglomeration of particles originally dispersed in a rubber latex.

Colligative properties: Properties of a solution that are dependent on the number of solute molecules present.

Combustible: Capable of burning.

Combustion: Exothermic reaction of a combustible substance with an oxidising agent, accompanied by flames or glowing and/or smoke.

Co-monomer: One of the two or more monomer species that polymerize to form a copolymer.

Compatabilization: Any physical or chemical action which results in stabilization, (prevention to separate) of PB.

Compatible of polymer blend: Term indicating commercially attractive PAB: usually system homogenous to the eye with enhanced physical properties.

Compound, standard: A control or reference compound prepared according to a prescribed formula and mixing procedure.

Compound: An intimate admixture of a polymer(s) with all the material necessary for finished article.

Conditioning: The subjection of a material to a stipulated treatment so that it will respond in a uniform way to subsequent testing or processing. Such treatment at specified temperature and humidity is generally specified in standard test method.

Conventional sulphur vulcanizing system: As applied to NR and isoprene and butadiene based synthetic rubbers, a vulcanizing system using a relatively high proportion of presulphur and producing at an optimum cure, a network in which the combined sulphur exists predominantly in the polysulphidic crosslinks and in non–crosslink chain modifications.

Copolymer: A polymer formed from 2 or more types of monomers.

Crazing: Fine cracks, which may extend in a network on or under the surface or through a layer of a plastic material.

Critical chain length: The minimum chain–length required for the entanglement of the polymer chains.

Crosslink: A chemical bond bridging one polymer chain to another.

Cross-linking: Applied to polymer molecules, the setting up of chemical links, between the molecular chains. When extensive, as in most thermosetting resins, cross–linking makes one infusible super molecule of all chains. Crosslinking can be achieved by irradiation with high–energy electron beams or by chemical cross–linking agents such as organic peroxides.

Crystalline melting point : The temperature at which the crystalline structure in a material is breakdown.

Crystallinity: Crystallinity defines the orderly arrangement and compactness of the molecular chains forming the polymer. Normally can be attributed to the formation of solid crystals having a definite geometric form.

Cure: To change the physical properties of a material, by chemical reaction, which may be condensation polymerization, or vulcanization; usually accomplished by the action of heat and catalysts, alone or in combination with or without pressure.

Deflection temperature: The temperature at which a test specimen deflects/deforms to a specified distance at a given load. (Heat deflection temperature is described in ASTM D648 for plastics under flexural load).

Dielectric strength: The minimum required voltage at which the insulating material fails to resist the electric current is known as dielectric strength. The voltage required for such breakdowns is termed as breakdown voltage. Dielectric strength is measured in volts/mil.

Dimensional stability: Ability of plastic part to maintain its shape as moulded or fabricated.

Dispersing agent: A surface–active substance used to facilitate the suspension of solid compounding materials in a liquid medium and to stabilize the dispersion thereby produced.

Ductility: The ability of a material to deform plastically before fracturing.

Ebonite: A hard material made by sulphur vulcanization of rubber, in which the hardness is substantially obtained by the action of sulphur.

Elastic deformation: Deformation of a part under load, which is recoverable on removal of the load.

Elastic limit: The greatest stress which a material is capable of sustaining without any permanent strain remaining upon complete release of the stress.

Elasticity: The property of the material to recover to its original shape and size on removal of stress, which causes deformation. When stress is proportional to strain, the material is called ideal elastic.

Elastomer: A macromolecular material at room temperature returns rapidly to approximately its initial dimensions and shape after substantial deformation by a weak stress and release of stress.

Emulsifying agent: A surface-active substance used to facilitate the dispersion of an immiscible liquid compounding material in another liquid and to stabilize the emulsion thereby produced.

Engineering polyblends: Polymer alloy of polymer blend

exhibiting properties and/or containing engineering polymers, at least one component of the blends.

Environmental stress cracking: Cracking or surface crazing which occurs on the products of thermoplastic material in contact with the chemicals (environments) when stressed.

Epoxy plastics: These are thermoplastic or thermosetting plastics containing ether or hydroxy alkyl repeating units, or both resulting from the ring opening reactions of lower molecular weight polyfunctional oxirane resin or compounds with catalyst or with various polyfunctional acidic or basic co–reactants.

Extender: An organic material used to augment the polymer in a compound.

FDA : Food and Drug Administration, USA

Fibres: A fibre is a thread or thread–like structure composed of strings or filaments of linear macromolecules that are cross–linked in such a manner so as to give rise to an assemblage of molecules having a high ratio of length to width.

Filler: A relatively inert material added to a plastic to modify its strength, permanence, working properties or other qualities or to reduce costs.

Film: In plastics, an optional term for sheeting, having a nominal thickness not greater than 0.25 mm.

Flame retardant: Substance that markedly retards the propagation of a flame.

Flammability: Ability of material or product to burn with a flame under specified test conditions.

Flash point: Minimum temperature at which, under specified test conditions, a material gives off sufficient flammable gas to produce a flash on applying the ignition source.

Functionality: The number of reactive groups in a molecule.

Gel point: The stage at which a liquid begins to exhibit pseudo–elastic properties.

Glass transition : The reversible change in an amorphous polymer or in amorphous regions of a partially crystalline polymer from (or to) a viscous or rubbery condition to (or from) a hard and relatively brittle one.

Glass transition temperature (T_g) : The approximate mid point of the temperature range over which the glass transition takes place is the transition temperature between glassy phase to rubbery phase.

Hardener: A chemical substance which is added to a thermosetting resin for the purpose of curing or hardening. Amines and acid anhydrides are hardeners for epoxy resins. Such chemicals are part of the chemical reaction and a part of the chemical composition of the cured resin.

Hydrophillic: Water loving or the material which absorbs moisture/water.

Hydrophobic: The material which does not have the tendency to absorb moisture/water.

Ignition temperature/point: Minimum temperature of a material at which sustained combustion can be initiated under specified test conditions.

Inhibitor: An additive that reacts with a chain–forming radical to produce non–radical products or radicals of low reactivity, incapable of adding fresh monomer units.

ISO : International Standards Organization

Isotactic: A chain of unsymmetrical molecules combined head-to-tail with their pendent groups occupying the same relative position in space along the chain.

Kinetic Chain-length: It is defined as the average number of monomer molecules contained per radical which initiates a polymer chain.

Melt flow index (MFI): The rate of extrusion of the thermoplastic resin in molten condition under specified temperature and pressure through a specified die is defined as melt flow index (MFI). It is expected in weight of extrudate obtained in 10 min (g/10 min).

Molecular weight distribution (MWD): The relative amount of polymers of different molecular weights that comprise a given specimen of a polymer.

Molecular weight: The sum of the atomic weights of all atoms in a molecule. In high polymers, the molecular weight of individual molecules varies widely, therefore, they are expressed as weight average or number average molecular weight.

Monomer (*mono* = single, *mer* = unit): A monomer is a relatively simple compound that can react to form a polymer (multiunit) by combination with itself or with other similar molecules or compounds, e.g. styrene, acrylate, propylene, vinyl chloride, etc.

Orientation: The alignment of molecules in the direction of flow. This can be accomplished by cold drawing or stretching during fabrication. When stretched in one direction, it is called monoaxial orientation and the stretching in two perpendicular directions provides *biaxial orientation.*

Plastic(s): A material that contains as an essential ingredient one or more organic polymeric substances of large molecular weight, is solid in its finished state and at some stage in its manufacture or processing into finished articles, can be shaped by flow.

Plasticized PVC (PPVC): PVC formulation containing softener and/or plasticizer.

Plastification: Softening of plastic material through heat and/or shear work.

Polymer (*Poly* = many, *mer* = unit): A polymer is high molecular weight organic compound whose structure can be represented by a repeated monomer unit. OR A substance consisting of molecules characterized by the repetition (neglecting ends, branch, junctions and other minor irregularities) of one or more types of monomeric units (IUPAC) (1971).

Polymer alloys: Polymer blends with modified interfacial properties and /or morphology.

Polymer blends: Mixture of chemically different polymers and/or copolymers. Miscible PB, polymer blend behaving as a single–phase system down to molecular level of dispersion.

Polymerization: A chemical reaction in which the molecules of monomers are linked together to form polymers.

Refractive index: The ratio of the velocity of light in a vacuum to its velocity in a substance. Also the ratio of the sine of the angle of incidence to the sine of the angle of refraction.

Reinforced plastics: Plastics with some strength properties greatly superior to those of the base resin, resulting from the presence of high strength fillers embedded in the composition. The reinforcing fillers can be fibres, fabrics, mats, etc.

Relative humidity (RH): The ratio of the quantity of water vapour present in the air to the quantity which would saturate it at any given temperature.

Resistivity: The ability of a material to resist passage of electrical current either through its bulk or on a surface.

Rubber cement: An adhesive that is a dispersion or solution of raw or compounded rubber or both, in a suitable liquid.

Sample: A small portion or part of a material or product intended to be representative of the whole.

Smoke: Visible suspension of liquid and/or solid particles in gases resulting from combustion.

Soot: Finely divided particles mainly carbon, produced and deposited during the incomplete combustion of organic material.

Specific gravity: The ratio of the weight of the given volume of a material to that of an equal volume of water at a stated temperature.

Specification: A detailed description of requirements, dimensions, materials etc.

Specimen: A piece or portion of a sample used for testing.

SPI : Society of Plastic Industry

Standard: Something established for use as a rule or as a basis of comparison in measuring or judging the quantity, value etc.

Tacticity: The arrangement of pendant groups in space.

Thermal conductivity: Ability of a material to conduct heat, physical constant for quantity of heat that passes through unit cube of a substance in unit time when difference in temperature of two faces is 1°C (ASTM C177).

Thermoplastics: A plastic that can be softened repeatedly by heating and hardened by cooling through a temperature range, characteristic of the plastics, and that in the softened state can be shaped by flow into articles by moulding or extrusion.

Thermoset: A plastic that, after having been cured by heat or other means, is substantially infusible or insoluble.

Theta temperature: A temperature at which a polymer of infinite molecular weight starts to precipitate from a solution.

Thixotropic: Describing materials that are gel-like at rest but fluid when agitated.

Unplasticized PVC: Also known as *rigid PVC* wherein softeners and/or plasticisers are either absent or loaded in such quantity that the material does not loose its rigidity.

Vicat softening temperature: the temperature at which a flat end needle of 1 mm^2 circular cross section will penetrate a thermoplastic specimen to a depth of 1 mm under specified load using a selected uniform rate of temperature rise.

Visco-elastic: A characteristic mechanical behaviour of some materials showing a combined behaviour, both viscous and elastic.

Viscosity: A measure of resistance of flow due to internal friction when one layer of fluid is caused to move in relationship to another layer.

Volume resistivity: Electrical resistance between the two faces of a 1 cm cube of the material. It is measured under specified conditions using a direct current potential. It is commonly expressed in ohm-cm.

Water vapour permeability: The time-rate of water vapour transmission through unit area of flat material of unit thickness induced by unit vapour pressure difference between two specific surfaces, under specified temperature and humidity conditions.

Water vapour transmission rate (WVTR): The steady water vapour flow in unit time through unit area of a body, normal to specific parallel surfaces, under specific conditions of temperature and humidity at each surface.

Ziegler-Natta catalyst: A catalyst with the composition $TiCl_4$ $-AlR_3$, obtained from titanium tetrachloride and aluminum trialkyl.

APPENDIX III

Conversion Tables

SI units and conversion factors

Physical quantity	*Name of unit in SI*	*Symbol for SI Unit*	*Definition of SI unit*
Length	Meter	m	Basic Unit
Mass	Kilogram	kg	Basic Unit
Time	Second	s	Basic Unit
Force	Newton	N	$kg\ m\ s^{-2}$ (= $J\ m^{-1}$)
Pressurc	Pascal	Pa	$kg\ m^{-1}\ s^{-2}$ (= $N\ m^{-2}$)
Energy	Joule	J	$kg\ m^2\ s^{-2}$
Power	Watt	W	$kg\ m^2\ s^{-3}$ (= $J\ s^{-1}$)

Conversion of Units

Physical quantity	*Customary unit*	*SI unit*	*To convert from customary unit to SI unit multiply by*
Length	in	m	2.54×10^{-2}
Mass	lb	kg	4.536×10^{-1}
Force	dyne	N	1×10^{-5}
	kg	N	9.80665
	lbf	N	4.44822
Pressure	dyne/cm^2	Pa or N m^{-2}	1×10^{-1}
	atm.	Pa or N m^{-2}	1.01325×10^{5}
	mm Hg	Pa or N m^{-2}	1.33322×10^{2}
	lbf/in.^{2} or psi	Pa or N m^{-2}	6.89476×10^{3}
Energy	erg	J	1×10^{7}
	Btu	J	1.055056×10^{3}
	ft-lbf	J	1.33582
	cal	J	4.187
	eV	J	1.602×10^{-19}
	Newton-meter	J	1
Area	in^2	m^2	6.4516×10^{-4}
	ft^2	m^2	9.290304×10^{-2}
Density	lb/ft^3	kg m^{-3}	16.02
Viscosity	poise (dynes/cm^2)	kg m^{-1} s^{-1} N s m^{-2}	1×10^{-1}
Viscosity, kinematic	Stoke (cm^2/s)	m^2 s^{-1}	10^{-4}
Surface tension	lbf/ft	N m^{-1}	14.59
	dyne/cm	N m^{-1}	10^{-3}
Stress	dyn/cm^2	Pa	0.1
	N/m^2	Pa	1
	J/m^3	Pa	1
	mm/Hg	Pa	133.3
	PSI	Pa	6.89°
	bar	Pa	1×10^{5}
	MPa	PSI	145

Reader's Notes

Reader's Notes

Reader's Notes